Advanced Laser Diode Reliability

**Durability, Robustness and Reliability
of Photonic Devices Set**

coordinated by
Yannick Deshayes

Advanced Laser Diode Reliability

Edited by

Massimo Vanzi
Laurent Béchou
Mitsuo Fukuda
Giovanna Mura

ELSEVIER

First published 2020 in Great Britain and the United States by ISTE Press Ltd and Elsevier Ltd

ISTE Press Ltd
27-37 St George's Road
London SW19 4EU
UK

www.iste.co.uk

Elsevier Ltd
The Boulevard, Langford Lane
Kidlington, Oxford, OX5 1GB
UK

www.elsevier.com

Notices

Knowledge and best practice in this field are constantly changing. As new research and experience broaden our understanding, changes in research methods, professional practices, or medical treatment may become necessary.

Practitioners and researchers must always rely on their own experience and knowledge in evaluating and using any information, methods, compounds, or experiments described herein. In using such information or methods they should be mindful of their own safety and the safety of others, including parties for whom they have a professional responsibility.

To the fullest extent of the law, neither the Publisher nor the authors, contributors, or editors, assume any liability for any injury and/or damage to persons or property as a matter of products liability, negligence or otherwise, or from any use or operation of any methods, products, instructions, or ideas contained in the material herein.

For information on all our publications visit our website at http://store.elsevier.com/

British Library Cataloguing-in-Publication Data
A CIP record for this book is available from the British Library
Library of Congress Cataloging in Publication Data
A catalog record for this book is available from the Library of Congress
ISBN 978-1-78548-154-3

Printed and bound in the UK and US

Contents

Introduction ix
Laurent BÉCHOU, Mitsuo FUKUDA, Giovanna MURA and Massimo VANZI

Chapter 1. Laser Diode Reliability 1
Mitsuo FUKUDA and Giovanna MURA

1.1. Laser diodes and application fields 1
1.2. Basic degradation mechanisms in laser diodes 3
  1.2.1. Device structure. 3
  1.2.2. Main degradation mechanisms 4
1.3. Degradation and reliability. 15
  1.3.1. Reliability in terrestrial and submarine systems 15
  1.3.2. Reliability in space application fields 17
  1.3.3. Display, lighting and storage 22
1.4. Physical analysis of degraded lasers 25
  1.4.1. Catastrophic optical damage 26
  1.4.2. Electrostatic discharge. 36
1.5. References 42

**Chapter 2. Multi-Component Model for Semiconductor
Laser Degradation** 51
Samuel K.K. LAM and Daniel T. CASSIDY

2.1. Introduction. 51
2.2. The physical explanation for saturable degradation 52
2.3. Rate equation for saturable defect population. 54
2.4. Saturable laser degradation by single defect population. 58
2.5. Multicomponent model for degradation dynamics 61
2.6. Annealing effect 63

2.7. Guide to MCM applications   69
2.8. Summary   74
2.9. References   75

**Chapter 3. Reliability of Laser Diodes for High-rate
Optical Communications – A Monte Carlo-based
Method to Predict Lifetime Distributions and
Failure Rates in Operating Conditions**   79
Laurent MENDIZABAL, Frédéric VERDIER, Yannick DESHAYES, Yves OUSTEN,
Yves DANTO and Laurent BÉCHOU

3.1. Introduction.   79
3.2. Methodology description   83
    3.2.1. Application context.   83
    3.2.2. Monte Carlo random sampling   86
    3.2.3. Brief review on random and pseudo-random
    numbers   87
    3.2.4. Direct Monte Carlo method: non-rectangular
    distribution   88
3.3. Description of the experimental approach.   89
    3.3.1. Choice of the correlation law   89
    3.3.2. Calculation of failure times and failure rates.   91
    3.3.3. Application to DFB laser diodes emitting at 1550 nm   92
3.4. Robustness analysis of the proposed method   96
    3.4.1. Validation of the method: analytical approach.   96
    3.4.2. Robustness analysis of the statistical random draws   103
3.5. Experimental investigations   117
    3.5.1. Single parameter estimation: drift of the bias current
    during aging tests.   117
    3.5.2. Multiple parameters estimation: drifts of threshold
    current and optical efficiency.   118
    3.5.3. Summary.   125
3.6. Toward multi-components physical models.   126
    3.6.1. Chuang's model: relationship between I_{th} and
    intrinsic density of defects   127
    3.6.2. Lam's model: growth of defects   130
3.7. Conclusion   132
3.8. References   134

Chapter 4. Laser Diode Characteristics.   139
Massimo VANZI

4.1. Introduction.   139
4.2. Energies and densities   143

4.3. Rates and balances. 148
    4.3.1. Equilibrium 148
    4.3.2. Quasi-equilibrium: the rate equation 150
4.4. Photon density 154
4.5. Spectral gain 158
4.6. Integral quantities: current I_{ph} and total power P_{OUT}. 164
4.7. Non-radiative current I_{nr}, threshold current I_{th} and the
light–current curve 167
    4.7.1. Non-radiative current I_{nr} and the total current I 168
    4.7.2. Gain–current relationship 171
    4.7.3. Measured optical power P_{OUT}. 171
4.8. Resistive effects 173
    4.8.1. Series resistance R_S. 173
    4.8.2. Side ohmic paths: current confinement 174
    4.8.3. Leakage paths 175
    4.8.4. Total current and its components. 175
4.9. Non-idealities 176
4.10. Appendix A: the anomaly ε 181
4.11. Appendix B: optical losses 185
4.12. Appendix C: a continuity equation for photons 187
4.13. Appendix D: the integral of the spectral function. 191
4.14. Appendix E: the lateral current I_W. 194
4.15. Appendix F: the gain–current relationship and
its comparison with the literature. 198
    4.15.1. Gain equations in the literature 198
    4.15.2. Comparison 201
4.16. References. 204

Chapter 5. Laser Diode DC Measurement Protocols 207
Massimo VANZI, Giovanna MURA, Laurent BÉCHOU,
Giulia MARCELLO and Valerio Sanna VALLE

5.1. The standard *LIV* curve: voltage or current driving 207
5.2. Voltage driving: the *logLIV* plot 208
5.3. Removing bad data: current compliance and
ambient photocurrent. 208
5.4. Calculating internal threshold voltage V_{th} and series resistance
R_S: the *logLIV* curves with respect to the internal voltage V. 209
5.5. *Canonical logLIV*: upscaling P_{OUT} to I_{ph}. 211
    5.5.1. Calculating I_{th} and reconstructing P_{TOT} and I_{ph}
from P_{OUT}: quantum efficiency η_q. 211

5.6. Subthreshold Shockley parameters for I_{ph}: saturation
current I_{ph0}, ideality factor n and quantum efficiency η_q   213
5.7. Lateral current I_W: current confinement   214
5.8. Transparency voltage V_{tr} for peak emission and zero-loss
threshold current I_{th0}   215
    5.8.1. The loss-absorption ratio α_T/g_m.   215
    5.8.2. Application to diagnostics.   216
5.9. Graphical interpretation of changes in DC characteristics   221
5.10. Gain measurements   225
    5.10.1. Non-resonating optical cavities.   225
    5.10.2. Fabry–Perot and DFB cavities   228
5.11. Appendix: a quick recall of the least squares method
for simple cases   231
5.12. References.   233

Introduction to Appendix   235
Massimo VANZI

Appendix. The Rules of the Rue Morgue   237
Massimo VANZI

List of Authors   255

Index   257

Introduction

There is a subtle line separating the impact of reliability, in the world of technology, between a reactive and a proactive role. Detecting and counting failures as well as predicting their occurrence on the basis of the accumulated data are the reactive aspects, driven by statistics (i.e. by measuring occurrences) and leading to important parameters such as lifetime or failure rate prediction. It is a crucial role of reliability that allows the design of suitably redundant systems to mitigate the impact of critical components on the overall operational life of an equipment.

The proactive side of reliability is completely different. It leads to technological improvement and is based on physics of electron devices first, and then particularly on those physics that aim to understand nature and kinetics of aging mechanisms. It is the discovery and explanation of the physical causes of failures that address the corrective actions to be undertaken in terms of the design and process of devices. Proaction has been an exciting but quite challenging game between the inventors of new technologies/devices and the reliability engineers. It improved the lifetime of electronic equipment so much that, if in the early '50s of the past century it was common to have 50% of apparatuses as radars under maintenance, in a few decades electron devices approached failure rates as low as 1 FIT, a new failure unit that represents the number of failures that can be expected in one billion (10^9) device-hours of operation.

This improvement action included the consideration of a variety of known and expected risks, but also the discovery of unexpected phenomena that in turn changed the manufacturing process. It was the case of latch-up in complementary metal-oxide-semionductor (CMOS) devices that prompted specific layout rules at design level; it occurred with the discovery of the many metal/metal or metal/semiconductor interface interactions, or with corrosion phenomena able to dissolve

---

Introduction written by Laurent Béchou, Mitsuo Fukuda, Giovanna Mura and Massimo Vanzi.

gold; again, it happened with microscopic mass transport such as electromigration. The reliability investigator needs to be an effective and proven specialist of both physics and engineering, and it is often his/her experience that will complement his/her skill and knowledge, without forgetting a little bit of luck.

Speaking of laser diodes, we enter a world that is even more special than solid-state electronics. Light interaction with matter introduces new physical phenomena in both operation and degradation of such devices. Here, skill, patient investigation, some luck and lots of experience in the field are the required tools for facing the new challenges that a still evolving technology continuously proposes.

This book is the coordinated effort of four teams of researchers, distributed over three continents. Adding up the years of experience of each team, more than one century of study is discussed in this book.

This book is not a textbook or a collection of separate contributions: several chapters have been written by mixed teams that have been discussing the topic over the many past years.

– Chapter 1 deals with degradation failure mechanisms, that is the failure physics of laser diodes.

– Chapter 2 faces the challenging problem of modeling the effect of multiple interacting mechanisms.

– Chapter 3 anchors predictive statistics to experimental data in the extreme case of high reliability devices.

– Chapter 4 proposes a model for reading the DC characteristics of a laser diode in terms of physical quantities relevant to degradation physics.

– Chapter 5 gives a summary of a set of procedures for measuring all parameters.

– The Appendix is a reprint of an old paper, not easily available, illustrating logics of failure analysis.

The editors, on behalf of all authors, hope that this book will become a valuable tool for reading the performances and possible degradations of laser diodes and for applying some practical innovative procedures for their future analysis in the field of reliability.

1

Laser Diode Reliability

1.1. Laser diodes and application fields

Laser diodes are the main optical source in various optical fiber communication systems and indispensable for our daily lives. The wavelength of the laser diodes are set at 1300 nm or 1550 nm for most communication systems, including submarine systems corresponding to the low loss window of silica fiber. Laser diodes used in such wavelength bands are usually InGaAsP/InP lasers. Wavelength in 850-nm band is also used for short-range communication, such as intraoffice systems. The laser diodes used for the wavelengths are AlGaAs/GaAs laser diodes. In these systems, high reliability is required for those laser diodes since the communication systems are serious lifelines, and the operating conditions have been severe year by year. In those applications for communication systems, high-speed operation for trunk systems and long-term stable wavelength operation for wavelength division multiplexing (WDM) systems are required. For subscriber systems, high and wide operating temperature are required, corresponding to the environmental conditions of applications. In addition to those applications, some laser diodes have begun to be used in space for satellite communications and sensing systems. The environmental conditions of space applications are quite different from those of terrestrial applications such as land and submarine communication systems. Compared with the operating conditions in terrestrial systems, there are large variations in the ambient temperature in space, and laser diodes are possibly exposed to severe irradiation by high-energy particles and ultraviolet rays. These environmental factors are inherent features in space applications and have been added to the reliability issues of terrestrial applications. The basic reliability of laser diodes used in the communication systems is usually determined by optical output characteristics. The stability of lasing wavelength and spectral linewidth are also important in these applications.

Chapter written by Mitsuo FUKUDA and Giovanna MURA.

For application in optical sensing systems, narrow spectral linewidth and wavelength tunability are important characteristics. Laser diodes are operated under DC current in combination with temperature change or under low-frequency AC current such as saw-shaped current to scan the wavelength. By scanning wavelength, absorption spectra of gases are monitored. The limiting factor of those applications is basically optical output power. The scanning temperature range can be widened if the output power of laser diodes is large because of a margin of operating current. From the reliability aspect, this problem is similar to the case of laser diode stability under a constant output power operation. The degradation behaviors used in sensing systems are similar to those of laser diodes used in communication systems if the laser diodes have pn-junctions. Quantum cascade lasers, which have no pn-junctions and are not diodes, show different behaviors of degradation because no radiative recombination occurs in the light emitting mechanisms. The quantum cascade lasers are highly reliable when compared with laser diodes with pn-junctions.

In consumer electronics, laser diodes have widely spread to various kinds of equipment, typically compact disk (CD) and digital versatile/video disk (DVD) systems. AlGaAs/GaAs laser diodes lasing at 780 nm are used in CD systems, and AlInGaP/In GaPlaser diodes lasing at 650 nm are used in DVD systems. AlGaAs/GaAs laser diodes are also used in printers, vending machines, etc. InGaN/GaN laser diodes emitting blue light are important optical sources in optical disk systems to increase a storage capacity and display equipment.

As described above, laser diodes are key components in many applications. Several causes of degradation have been reported in correspondence with the structure and material of laser diode and operating conditions. Degradation gradually occurs even in recent systems and equipment. In space, radiation damage is a cause of the degradation of laser diodes. Radiation damage can be reduced by packaging and shielding laser diodes from high-energy particles and ultraviolet rays. The longest lifetime of laser diodes in space, therefore, corresponds to the lifetime in terrestrial systems.

This chapter reviews the main degradation mechanisms and reliability issues of laser diodes mainly used in communication systems because higher reliability is usually required of laser diodes used in communication systems. After reviewing the causes of degradation and changes in device characteristics of laser diodes used in terrestrial and submarine systems, the discussion focuses on the performance degradation of laser diodes used for many systems. In the final part, the sample preparation through thick lamella analyzed by using the scanning transmission electron microscopy is proved to be successful in the analysis of several degraded lasers.

1.2. Basic degradation mechanisms in laser diodes

1.2.1. *Device structure*

The basic structure of a laser diode is planar type with a double heterostructure. Two typical structures are shown in Figure 1.1. Figure 1.1(a) shows an early type of double heterostructure laser diode developed in the initial stage of laser diode history (Fukuda 1991). A ridge-waveguide laser diode is shown in Figure 1.1(b). The lasing optical field of the device in Figure 1.1(b) is confined within a relatively narrow region under the ridge-waveguide. A laser diode of this type includes 980 nm band strained InGaAs/GaAs laser diodes for pumping Er-doped fiber amplifiers and optical sources of fiber communication systems. In most laser diodes used in optical fiber communication systems, the lasing optical field is confined in a very narrow region to reduce the lasing threshold current and operating current and to stabilize the transverse lasing mode. Various buried heterostructure (BH) laser diodes have been developed to confine the optical field. Two typical BH laser diode structures are shown in Figure 1.2. A waveguide is formed in the light-emitting regions. This waveguide structure is fabricated by physically/chemically etching the epitaxial layers down to a mesa structure, and the mesa is then buried by semiconductor layers or dielectric films. The main optical sources in current systems employing single mode fiber are distributed feedback (DFB) laser diodes (see Figure 1.3). To further improve the performance of the laser diodes in optical communication systems, the electroabsorption-type modulator-integrated DFB laser diode (EA-DFB) has been developed, and this type of laser diode has become an important light source in dense WDM (dWDM) systems. The laser section operates under DC bias, and the lasing light is modulated in the modulator section. The degradation modes and reliability of the laser section are basically the same as in a solitary laser diode used in optical fiber communication systems.

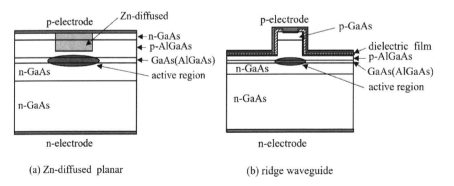

Figure 1.1. *Gain guiding-type laser diodes*

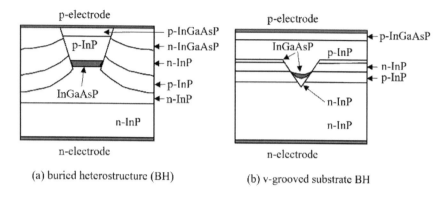

(a) buried heterostructure (BH)

(b) v-grooved substrate BH

Figure 1.2. *Refractive index guiding laser diodes*

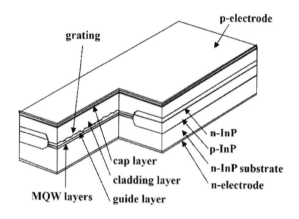

Figure 1.3. *Distributed feedback laser diode (buried heterostructure type)*

1.2.2. *Main degradation mechanisms*

Figure 1.4 summarizes the main parts, at which degradation tends to occur, in BH laser diodes (Fukuda 2000, 2007). They include growth of dislocation networks in the inner region, defect increase at the BH interface, photo-enhanced oxidation of the facets, reactions between the electrode and the semiconductor, and solder instability at the bonding part. The causes of the degradations have been eliminated or nearly completely suppressed by employing suitable device structures and materials, improving device processes and carrying out burn-in/screening tests.

Figure 1.5 summarizes the main degradations and the characteristic changes of 1300- and 1550-nm band laser diodes. The main cause limiting the lifetime of laser diodes used in optical fiber communication systems is degradation at the BH interface.

The schematic diagram of the degradation in BH interfaces is shown in Figure 1.6. The increase in nonradiative leak current at the surface/interface of the pn-junction is the main cause of the BH degradation. During the degradation, defects increase at the interface between the active region and the burying layer. In severe cases, these defects form dislocation networks even in an active layer composed of InGaAsP, although the material system is usually insensitive to defects. The injected current lost by nonradiative recombination gradually increases when defects increase at the BH interface.

In this situation, the threshold current increases but the slope efficiency remains nearly constant in 980-, 1300- and 1550-nm band laser diodes, as shown in Figure 1.7 because the carrier lifetime of the stimulated emissions is shorter than that of spontaneous and nonradiative recombination. The increase in defect density and the growth of the dislocation networks are generated in, or in the vicinity of, the active region during operation. The degradation mechanisms described above are already discussed in Fukuda (1991, 2000, 2007).

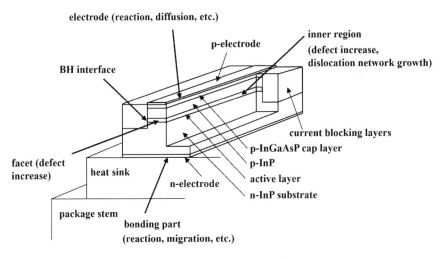

Figure 1.4. *Degradation mechanisms in LEDs and laser diodes*

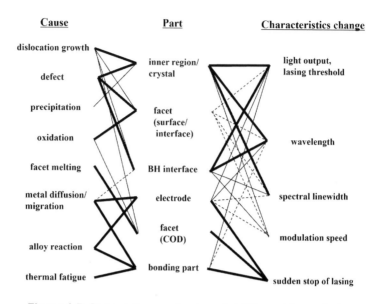

Figure 1.5. *Degradation mechanisms in LEDs and laser diodes*

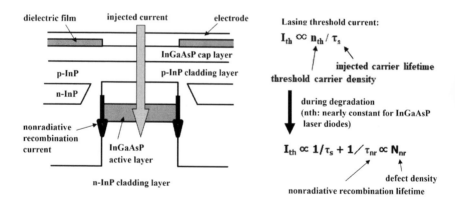

Figure 1.6. *Degradation mechanism of buried heterostructure InGaAsP/InP laser diode and change in characteristics*

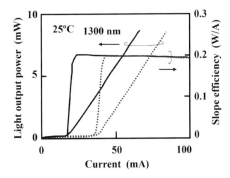

Figure 1.7. *Change in threshold current and slope efficiency as a result of BH interface degradation*

The degradations are shown in Figure 1.5 and various degradation mechanisms deteriorate the performances of laser diodes. In the following sections, the characteristic changes caused by the degradation are discussed in detail.

1.2.2.1. *Change in wavelength during degradation*

The lasing wavelength of laser diodes is basically determined with the band gap of the active layer in Fabry–Perot types and the refractive index in DFB and distributed Bragg reflector (DBR) types, as shown in Figure 1.8 (Fukuda *et al.* 2010). The laser diodes with Fabry–Perot type cavity lase at the wavelength corresponding to the bandgap energy of the active layer due to the transition mechanism. When the temperature of laser diodes increases, the active layer (lattice spacing) expands with heat. This results in the reduction of the bandgap energy of the active layer, and the lasing wavelength increases corresponding to the expansion of the lattice spacing. If the temperature of laser diodes decreases, reverse procedure occurs and the lasing wavelength decreases. In addition, the band-filling effect strongly influences the emitting wavelength of laser diodes before lasing (LED mode). The carriers injected in the active layer are successively packed into the energy level from the conduction band minimum for electrons and the valence band maximum for holes. Before lasing (LED mode), the emitting wavelength gradually shortens as the injected current increases, as shown in Figure 1.9. The wavelength shortening suddenly stops at the lasing threshold because the light emission mechanism changes from spontaneous emission to stimulated emission. Here, injected carriers are mostly converted to light because the lifetime of stimulated emission is more than one order of magnitude shorter than that of spontaneous emission, and the light is emitted to the outside of laser diodes as lasing light. At the lasing threshold, therefore, the injected carrier is nearly constant. After lasing, wavelength usually increases according to the Joule heating (see Figure 1.9).

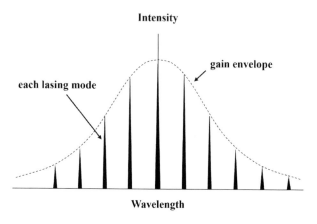

Figure 1.8. *Lasing wavelength of FP laser and DFB/DBR laser diodes. The gain envelope determines the shape of longitudinal modes in FP lasers, and a mode is selected with a grating structure in DFB and DBR laser diodes*

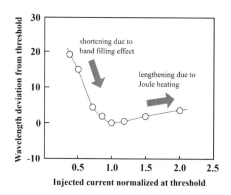

Figure 1.9. *Change in peak wavelength of a 1550 nm wavelength band FP-type MQW (multi quantum well) laser diode with 300 μm long cavity*

For DFB- or DBR-type laser diodes, the same mechanisms as phenomena in Fabry–Perot laser diodes govern the emitting wavelength before lasing. At around the lasing threshold, the grating pitch determines the lasing wavelength. The pitch of grating is a function of the refractive index, and the selected lasing wavelength changes with the value of the refractive index of the active layer. The refractive index corresponds to the square root of the relative dielectric constant, and is connected to atomic (or electronic) polarization. The polarization is influenced from the lattice spacing, and the refractive index changes with temperature and lasing (selected) wavelength shifts by the refractive-index change (see Figure 1.10).

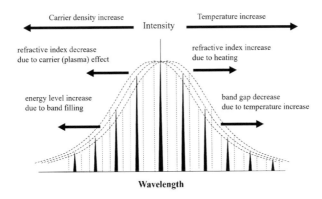

Figure 1.10. *Change in lasing wavelength of PF laser and DFB/DBR laser diodes*

In addition to these factors, transverse mode change in planar type and ridge waveguide type laser diodes influence the wavelength shortening or lengthening during degradation. For DFB-type laser diodes, especially with high coupling constant, kL, the change in optical intensity distribution along the cavity leads to wavelength change (jumping to another mode).

During degradation, the lasing wavelength changes according to the mechanisms described above, but we can analyze the behaviors of lasing wavelength based on the mechanisms described above. Actual examples are discussed in the following sections.

1.2.2.2. Wavelength change during degradation in laser diode chip

During degradation, threshold carrier density increases to compensate optical loss increase and injected carrier lifetime reduction caused by nonradiative recombination (Fukuda *et al.* 2010). Joule heating is also a serious problem under operation at a constant output power. These characteristic changes induce wavelength change as shown in Figure 1.10.

1) Wavelength shortening

The causes of lasing wavelength shortening are as follows:

i) the band-filling effect for the Fabry–Perot type: introduced by lasing threshold current increase;

ii) refractive index reduction for the DFB type: introduced by threshold carrier density increase.

2) Wavelength lengthening

The causes of lasing wavelength shortening are as follows:

i) bandgap reduction for the Fabry–Perot type: introduced by Joule heating;

ii) refractive index increase for the DFB type: introduced by thermally expanding lattice spacing due to Joule heating.

1.2.2.3. *Wavelength change during degradation in the outside of laser diode chip*

Lasing wavelength is influenced by the materials used, device structure and operating conditions (Fukuda *et al.* 2010). These are originated from laser diode chip itself in most cases. Another important factor except for chip strongly influences the wavelength change. That is the degradation of the bonding part induced by soldering instability. Heat generated at the active layer quickly spreads to the laser diode chip, heat sink, submount, package stem, package, etc., as shown in Figure 1.11. If the bonding part, such as the interface between chip and heat sink, degrades, the heat conductance decreases at the interface. This results in temperature rise in chip, and lasing wavelength lengthening and output power reduction are generated. This type of degradation occurred at the interface between chip and heat sink or between heat sink and package stem during long-term operation, if low melting point solder (soft solder) was used to bond them. This interface degradation results in so-called thermal runaway under operation at a constant output power.

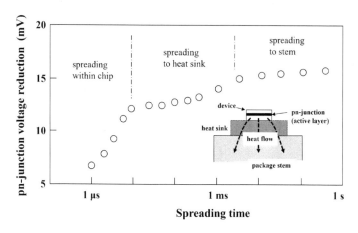

Figure 1.11. *Heat spreading from active layer to heat sink and stem under forward current of 100 mA. The sample is a 1300 nm wavelength band LED. The temperature is monitored as the reduction of pn-junction voltage at 1mA due to the heating. The change coefficient of the voltage is about 1.2 mV/deg*

1.2.2.4. Laser diode lifetime and wavelength change

The lifetime is usually defined at the time when one of the characteristics required for operation in equipment or systems passes over the specified value (Fukuda 2000, 2007). The lifetime is defined with operating current under a constant output power in most systems or equipment. In addition to the basic characteristics, a certain character is often important for applications.

For optical fiber communication systems, the stability of the lasing wavelength is very important in dense-WDM systems. In those systems, the change of the lasing wavelength within 0.2 nm is required for 20 years. A DFB laser diode in the early stage, the wavelength change, is much larger than that of recently developed laser diodes. However, a certain feedback system to maintain the lasing wavelength is needed for d-WDM applications even for the recent DFB laser diodes. The wavelength deviation during operation is corrected with a wavelength locker, where the wavelength change is automatically compensated with temperature controlling the laser diode by monitoring the change in the wavelength using the locker.

In addition, to stabilize the lasing wavelength and simplify the module structure, electroabsorption-type modulator-integrated DFB laser diodes were often used in WDM systems. The degradation rate is usually limited with the rate of the DFB laser section because the modulator section is reversely biased and the performance is quite similar to that of photodiode.

Electrically wavelength-tunable type laser diodes can be roughly divided into two types, the solitary-laser type and the arrayed type. The lifetime depends on their types or structures. For solitary types such as DBR laser diodes, the lasing wavelength is electrically scanned, although the contentious tunable range is not so wide because of mode hopping. Here, the wavelength is selected with the DBR section by changing the grating pitch (refractive index) due to current injection. The contentious tunable range is gradually narrowed under degradation (Mawatari et al. 1999). The degradation of the DBR section is also governed by the causes described in Figure 1.5. The wavelength-set point should be centered in the tunable range to expand the wavelength margin from the viewpoint of lifetime. The lifetime of arrayed-type lasers depends on the way they are used. If only one laser of the array is operated, the lifetime is the same one as a conventional DFB laser. If a few laser diodes in the array are operated at the same time, the lifetime as an arrayed-type laser diode is determined with a laser diode having the shortest lifetime.

1.2.2.5. Change in coherence during degradation

Coherence is the essential characteristics of laser diodes as well as the other lasers, and the most characteristics specific to lasers depend on the coherence. All kinds of degradation of laser diodes, therefore, strongly influence the coherence, although the way to influence them is different with the degradation mechanisms. The coherence of semiconductor lasers is inferior to that of the other lasers because of the small cavity size, large cavity loss, refractive index fluctuation of cavity, etc. However, the coherence of semiconductor lasers, especially laser diodes operating at a single mode, is greatly improved by employing quantum structure to the active layer. This coherence can be monitored with the lasing spectral line width determined by stimulated emission process in the cavity. The spectral linewidth is one of the important characteristics of laser diodes (and quantum cascade lasers) for coherent communication systems and sensing equipment.

1.2.2.6. Basic mechanisms of spectral linewidth change

The spectral linewidth, Δv, is an essential characteristic for single-mode laser diodes and can be given by multiplying the Schawlow–Townes equation by the constant term $(1 + \alpha^2)$ and expressed using the following equation:

$$\Delta v = (gn_{sp}/4\pi I_p) (1 + \alpha^2),\qquad\qquad [1.1]$$

where g is the gain coefficient, n_{sp} is the spontaneous emission coefficient, I_p is the number of photons, and α is the linewidth enhancement factor, which is governed by the fluctuation of refractive index and gain in the active layer under carrier injection. Equation [1.1] can be modified to the next equation using the absorption coefficient (internal loss), α_i, and mirror loss, α_m,

$$\Delta v \propto (\alpha_i + \alpha_m) \alpha_m (1 + \alpha^2).\qquad\qquad [1.2]$$

The coherence (spectral linewidth) of laser diodes lowers (broadens) through the factors in equation [1.2] under degradation. The gain for lasing and the linewidth enhancement factor gradually increase during degradation.

1.2.2.7. Spectral linewidth broadening

As described in section 1.2.2, various kinds of degradation mechanisms increase absorption coefficient and injected carrier density (Fukuda et al. 1993). The increase in carrier density resulted from the decrease in injected carrier lifetime caused by nonradiative recombination. These factors strongly influence the spectral linewidth of laser diodes. According to equation [1.1], the spectral linewidth gradually decreases as the lasing output power increases. The absorption coefficient directly broadens the linewidth as shown in equation [1.2] and Figure 1.12. At a certain point, the behavior deviates from the relationship expressed by equation [1.1], and quickly increases (rebroadens). After rebroadening, a single-mode laser diode cannot be used in equipment and systems requiring high coherence, such as laser diodes for a gas-sensing system, even though the submode is very small and cannot be monitored with a spectrometer, etc. This situation is detailed in Figure 1.12.

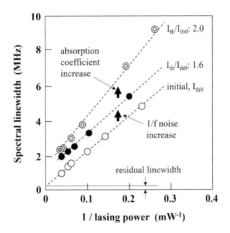

Figure 1.12. *Change in spectral linewidth during degradation. The sample is a 1550 nm wavelength band DFB laser diode having 900 µm long cavity*

1.2.2.8. *Residual linewidth increasing due to 1/f noise*

Another important factor influencing spectral linewidth is the so-called residual linewidth (Fukuda *et al.* 1993). According to equation [1.1], the spectral linewidth is zero if the lasing output power infinitely increases. Spectral linewidth never reaches the value of zero if the trend of the relationship between spectral linewidth and stimulated output power is extended to zero points on the horizontal axis, as shown in Figure 1.12. The deviation of the spectral linewidth from zero at the origin of the horizontal axis is called the residual linewidth.

This residual linewidth is caused by the increase in 1/f noise generated by nonradiative recombination at pn-junction and increases during degradation as shown in Figure 1.12. This residual linewidth is theoretically separated from equation [1.1] and an additional factor, and its influence on the coherent performance gradually decreases as the frequency used in systems is high.

1.2.2.9. *Change in modulation characteristics*

To increase modulation bandwidth of laser diodes, the active layers of laser diodes are often highly doped (Fukuda 1991). The increase in modulation bandwidth is resulted from the reduction of injected carrier lifetime. The similar carrier lifetime reduction is generated during degradation because crystal defects increase at the initial stage of degradation. However, output power gradually decreases during degradation and this deteriorates the modulation characteristics. The other degradation modes, such as the instability of the optical field distribution along the cavity, also deteriorate modulation characteristics of laser diodes.

The injected carrier lifetime is strongly influenced with defect density in the active layer and can generally be expressed with the next equation because each recombination is a parallel event:

$$1/\tau_s = 1/\tau_r + 1/\tau_{nr}, \qquad\qquad [1.3]$$

where τ_s is the injected carrier lifetime, τ_r is the radiative recombination lifetime, and τ_{nr} is the nonradiative recombination lifetime. The $1/\tau_{nr}$ is proportional to the density of defects within the laser cavity, and thus the nonradiative lifetime gradually decreases during degradation. The reduction of the nonradiative recombination lifetime results in an increase in relaxation oscillation frequency (or noise resonance frequency), which determines the upper limit of the modulation frequency of laser diodes. The relaxation oscillation frequency, f_r, is approximately expressed using the following equation:

$$f_r = A \, (1/\tau_s \, \tau_p)^{1/2}[(I - I_{th})/I_{th}]^{1/2}, \qquad\qquad [1.4]$$

where A is a constant, τ_p is the photon lifetime, I and I_{th} is the injected current and the threshold current, respectively. The relaxation oscillation frequency, therefore, increase through the reduction of τ_s during degradation. This situation is indicated in Figure 1.13. This injected carrier lifetime reduction is mainly generated in the initial stage of degradation. After the absorption coefficient starts to increase, optical output power decreases, and then electric bias cannot be set at the point of the defined output power. This finally deteriorates the direct modulation characteristics of laser diodes.

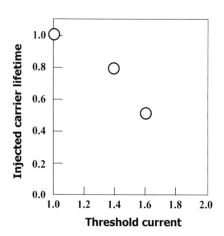

Figure 1.13. *Change in injected carrier lifetime during degradation on the assumption that the photon lifetime is constant. The threshold current on the horizontal axis and injected carrier lifetime on the vertical axis is normalized at the initial value, respectively. The sample is a 1550 nm wavelength band BH-type FP laser diode*

1.2.2.10. *Intensity noise increase*

The phenomena generated under degradation, such as injected carrier lifetime reduction and absorption coefficient increase, induce intensity noise increase (Fukuda 1991). The intensity noise deteriorates the signal-to-noise ratio (SNR) and directly influences the transmission quality in communication systems, especially analogue systems. As a measure of intensity noise, relative intensity noise (RIN) is usually used:

$$RIN = (N_p - N_n)/GR_L I_R^2 \Delta f, \qquad\qquad [1.5]$$

where N_p is the measured noise power, N_n is the noise power originated from the photodetector (shot noise) and amplifier (mainly thermal noise), G is the gain of the amplifier, R_L and I_R are the load resistance and the photocurrent of the photodetector, and Δf is the measuring frequency bandwidth. An example of the change in RIN is shown in Figure 1.14. During the increase in the threshold current, the RIN increases due to the noise power increase. The noise power is determined by the stability of the lasing situation and strongly influenced by gain and refractive index fluctuations, such as mode-competition noise, etc.

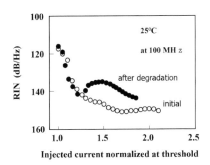

Figure 1.14. *Change in relative intensity noise (RIN) of 1300 nm wavelength band BH-type FP laser diode having 300 µm long cavity. The aging was carried out at a constant current of 200 mA and 100°C. The threshold current increased by 32% after degradation*

1.3. Degradation and reliability

1.3.1. *Reliability in terrestrial and submarine systems*

Laser diodes usually operate under constant output power in most communication systems. The change in operating current, therefore, exhibits different patterns when degradation is generated, as shown in Figure 1.15. The modes of current increase, sudden, rapid or gradual, are determined by the degraded part and by the cause of the degradation (Figures 1.4 and 1.5). These modes had been eliminated or suppressed to some extent in laser diodes used in traditional systems, but have reappeared in laser diodes in currently developed systems. In addition, new issues, such as precise stability of lasing wavelength, have been cited as reliability factors in the new systems. Consequently, the lifetimes and failure rates have not improved much over the years if one only compares the values. The median lifetimes required are usually about 10–20 years. These estimated lifetimes and reliability, of course, satisfy the requirements for communication systems (Fukuda 1991, 2000, 2007), although the required reliability is different with each communication system. In high-speed systems, such as ones for 10 Gb/s signal transmission, the stability of injected carrier lifetime for more than 10 years is important, and the wavelength stability of less than 0.2 nm for 20 years is requested in dWDM systems. Stability against ambient temperature changes ranging from −45 to +85°C for 10 years is an important characteristic for access systems. These reliability requirements are satisfied in each system.

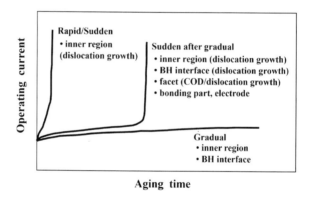

Figure 1.15. *Failure modes of laser diodes under constant output power operation*

Laser diodes currently used in such systems are mainly BH InGaAsP/InP and AlInGaAs/InP laser diodes (Nakahara *et al.* 2004; Ikoma *et al.* 2005). The performance is mainly deteriorated as a result of degradation at the BH interface and facet degradation, as described in the previous section. In particular, the nonradiative recombination current gradually increases at the BH interface in corresponding to the increase in crystal defects. Lasing characteristics mainly change as a result of injected-carrier lifetime shortening, which increases the nonradiative recombination current. In this degradation, the threshold current increases, the slope efficiency remains constant, the lasing wavelength remains nearly constant, and the modulation characteristics scarcely change. This degradation mode determines the device life and reliability of laser diodes in conventional as well as newly developed communication systems. The activation energy of the degradation at the BH interface was estimated to be around 0.5 eV.

1.3.2. *Reliability in space application fields*

The cause of degradation and the reliability of laser diodes were described above on the basis of common degradation patterns experienced in terrestrial and submarine optical fiber communication systems. Point defects in laser diodes currently used in terrestrial systems give rise to BH interface degradation and facet degradation. These degradations will be enhanced by radiation damage (point defects) in space, although almost all defects can be annealed. The possible highest reliability in space, therefore, corresponds to the reliability in terrestrial systems. The difference in reliability between space and terrestrial applications will be small by putting laser diodes meant for space applications in radiation-resistant packages and providing their equipment with temperature controls.

1.3.2.1. *Degradation mechanisms in space systems*

From the viewpoint of radiation damage, the reliability observed in terrestrial systems would be the best case in space because there is no radiation damage by protons, electrons and ultraviolet rays in terrestrial systems. The effects of the radiation damage are never included in the reliability estimation for terrestrial optical fiber communication systems. In contrast, laser diodes used in space are installed in certain equipment, such as satellites, and are protected from radiation damage. The structure of the package and equipment are important factors to protect laser diodes from radiation damage. Consequently, the reliability of laser diodes for space applications will be estimated as a function of the magnitude of radiation damage on the basis of the reliability of laser diodes in terrestrial optical fiber communication systems. In addition, severe ambient temperature conditions have to be considered for space applications.

The ambient temperature in terrestrial systems usually ranges from −40 to +85°C. Temperature conditions in space are much more severe than those in terrestrial applications, for instance, the temperature in space changes from −120°C to +120°C, although it can be artificially controlled to some extent.

1.3.2.2. Influence of radiation effect on degradation

The degradations associated with crystal (point) defects are dislocation network growth in the active layer, BH interface degradation and facet degradation, as indicated in Figure 1.5. These degradations are possibly enhanced in space because crystal defects are introduced under irradiation of protons, electrons and ultraviolet rays. The fluence densities strongly depend on the altitude of orbits set in a range from 600 to about 38,000 km. In higher orbits, laser diodes are exposed to a higher fluence density. The effect of radiation damage on the reliability of laser diodes has not been evaluated in long-term aging study. However, a few primitive tests have been done, and changes in device characteristics due to the effects of irradiation have been reported (Tan et al. 1996; Schone et al. 1997; Tan and Jagadish 1997; Johnston and Miyahira 2000; Johnston et al. 2001; Lippen 2002; Le Metayer et al. 2003; O'Neill et al. 2003; Johnston and Miyahira 2004; Khanne et al. 2004; Okada et al. 2008; Fukuda 2009). The most important radiation effect among them is the creation of crystal defects such as vacancies and interstitial atoms (displacement of atoms).

Proton irradiation (Tan et al. 1996; Schone et al. 1997; Tan and Jagadish 1997; Johnston and Miyahira 2000; Johnston et al. 2001; Le Metayer et al. 2003; Khanne et al. 2004; Okada et al. 2008), electron irradiation (Ando et al. 1984; Yamaguchi et al. 1984a, 1984b; O'Neill et al. 2003) and γ-ray irradiation (Yamaguchi et al. 1984) cause crystal defects in semiconductors. The amount of irradiation in these tests corresponded to operation times of a few decades in space. The defects generated under such irradiation are usually Frenkel–type defects, i.e. a vacancy and interstitial complex are created in semiconductor crystal. Such defects decrease the majority- and minority-carrier diffusion length and in turn the injected minority-carrier lifetime. Electrical resistivity also increases under irradiation, and the irradiation is sometimes used to form a region with high electrical resistance to separate devices and confine the current-flow region (Lippen 2002). The effect of these point defects on device reliability depends on the semiconductor material and its structure.

Just after point defects creation due to irradiation, the photoluminescence intensity decreased in compound semiconductors such as InP, GaAs and GaN (Ando et al. 1984; Yamaguchi et al. 1984a, 1984b, 1984c; Okada et al. 2008), the threshold current increased in laser diodes (Schone et al. 1997; Johnston and Miyahira 2000, 2004; Le Metayer et al. 2003), and the light-emission intensity decreased in LEDs (Johnston and Miyahira 2000; Khanne et al. 2004). The defects introduced by irradiation were greatly annealed by current injection under a forward bias, high temperature storage or light illumination (Yamaguchi et al. 1984a; Schone et al. 1997; Johnston and Miyahira 2000; Johnston et al. 2001; Le Metayer et al. 2003; Okada et al. 2008). During long-term operation in space, laser diodes composed of materials sensitive to such defects will exhibit degradation. In such materials, nonradiative recombination-enhanced defect motion occurs under current injection (Petroff and Hartman 1973, 1974; Lang and Kimerling 1974; Weeks *et al.* 1975; Petroff *et al.* 1980). Deep levels are created at around the mid gap in 850-nm band AlGaAs/GaAs laser diodes (and LEDs), and thus nonradiative recombination occurs at high rates. This recombination emits multiphonons, and the emitted energy enhances the defect motion and finally results in the growth of dislocation. In laser diodes composed of materials insensitive to crystal defects, such as InGaAsP/InP and AlInGaAs/InP, quick dislocation growth is rarely observed, although the degradation modes similar to those of AlGaAs/GaAs laser diodes are observed (Weeks *et al.* 1975; Petroff *et al.* 1980).

Defects introduced in space tend to be annealed under a light-emitting operation as shown in Figures 1.16(a) and (b). A small percentage of the defects, however, remain and deteriorate device characteristics, if they couple with other defects such as impurities. The behavior of the remaining point defects will be the same as those observed in laser diodes used in terrestrial systems. The residual point defects would add to the original defects generated under current injection and enhance the degradation of the BH interface, facet and inner region of laser diodes during operation.

Schematic diagrams of estimated BH and facet degradation are shown in Figures 1.17(a) and (b).

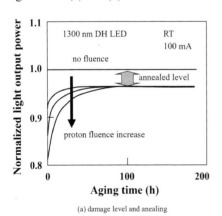
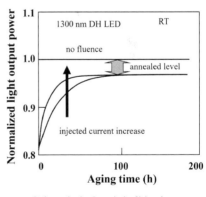

(a) damage level and anealing   (b) damage level and magnitude of injected current

Figure 1.16. *Radiation damage and its annealing effect as a function of aging time. The samples are 1300 nm LEDs*

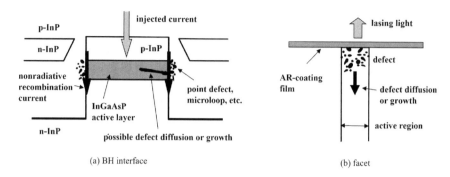

Figure 1.17. *Schematic diagram of BH (a) and facet (b) degradation. During operation in space, residual point defects (radiation damage) after annealing compound the defects generated during normal operation*

1.3.2.3. *Laser diode reliability in space*

In space, high-energy protons (or α-ray) and electrons (β-ray) have fluences ranging from 10^7 to 10^{15} cm^{-2} year^{-1}. Low energy fluences can be prevented with a cover glass but protons with energies higher than 5 MeV and electrons with energies higher than 0.2 MeV pass through cover glass in fluence that range from 10^7 to 10^{12} and from 10^7 to 10^{14} cm^{-2} year^{-1}, respectively. These radiation damages can be reduced with proper packaging and installation in the satellites. In addition, radiation damages will be greatly annealed during operation.

Under radiation damage, laser diodes can operate stably with a small amount of operating current increase. In this situation, the reliability might be comparable to that of laser diodes in terrestrial communication systems. If the defects introduced under radiation enhance the growth of line imperfections (dislocation networks), a possibility of rapid degradation after long-term stable operation will increase, as indicated with the "sudden after gradual" label in Figure 1.15. Here, two kinds of defect reactions have to be considered: recombination enhanced defect climb (REDC) and defect glide (REDG). Such defects were often observed in the initial developmental stage of laser diodes and correspond to <100> and <110> dark line defects (DLDs) when the active layer of laser diode is grown on the (100) crystal plane as shown in Figure 1.18. For REDC, laser diodes composed of materials insensitive to crystal defects, such as InGaAsP, InGaAs and AlInGaAs, will still operate stably because the growth of dislocation networks due to REDC is not very frequent. However, the REDC possibly degrades the BH interface and facet, as shown in Figure 1.17. Laser diodes composed of materials which are sensitive to crystal defects tend to generate the degradation caused by the growth of dislocation networks under current injection.

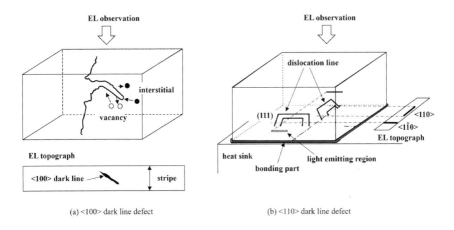

Figure 1.18. *Growth mechanisms of <100> (a) and <110> (b) dark line defects*

The other defect reaction, REDG, affects laser diodes composed of materials insensitive as well as sensitive to point defects. This type of dislocation network mainly grows as a result of mechanical shear stress from the electrode, bonding part, package stem, etc., and its growth rate depends on the magnitude of current injection and irradiation of high-energy particles such as electrons (Maeda and Takeuchi 1996; Tomiya *et al.* 2004). This type of degradation, which is the generation of <110> DLDs, might be likely in space.

The change in ambient temperature is severe in space. Without an effective control or countermeasure, this temperature change would adversely affect the reliability of laser diodes, as many reports have mentioned (Fukuda 2009). In particular, the Auger recombination effect tends to be intensified as the band gap of semiconductor material decreases (Thompson and Henshall 1980; Sugimura 1981), and thus the temperature characteristics of 1300-nm and 1550-nm band laser diodes are inferior to those of 980-nm band InGaAs/GaAs and 850-nm band AlGaAs/GaAs laser diodes. The characteristics of 1300-nm and 1550-nm band laser diodes significantly deteriorate at high ambient temperature. Here, the important factors influencing their performance and reliability are the carrier overflow from the heterobarrier in the active region of the laser diode and intervalence band absorption generated in the valence band of semiconductor material (Goodwin *et al.* 1975; Ettemberg *et al.* 1979; Asada and Suematsu 1983; Fukuda 1991). A temperature controlling is needed if these laser diodes are to be used for a long term in space.

1.3.3. *Display, lighting and storage*

Recently, many kinds of laser diodes have been used in various consumer electronics, such as 780-nm wavelength band AlGaAs/GaAs laser diodes in CD systems and 650-nm wavelength band AlInGaP/InGaP laser diodes in DVD systems. Reliability of the laser diodes used in such new fields can also be discussed on the basis of failure physics clarified for the laser diodes used in fiber optic communication systems.

1.3.3.1. *Discussion of laser diodes composed of wide bandgap materials*

InGaN/GaN laser diodes emitting blue light are important optical sources in display and lighting applications. The degradation modes and long-term stability of InGaN/GaN laser diodes can be discussed to some extent, although their reliability tests are not performed so much when compared with those of laser diodes used in communication systems.

The growth rate of a dislocation network, which is so-called <100> direction dislocation, was high in an AlGaAs/GaAs material system. This resulted from the high rate of the nonradiative recombination via deep levels in the bandgap. The nonradiative recombination emits multiphonons and enhances the defect motion. In an InGaAsP/InP material system, the dislocation network has rarely been observed because no deep level is located within its bandgap, and only little power is emitted to move the defects related to dislocation network. In an InGaN/GaN system, host atoms are tightly bonded because of the wide gap semiconductor, and the emitted energy might be too small to interact between the dislocation and defects such as interstitial atoms and vacancies. Under large mechanical stress, the <110> direction dislocation network is introduced in InGaAsP/InP as well as AlGaAs/GaAs laser diodes during operation. The growth rate of this dislocation is influenced by nonradiative recombination and the bonging strength of atoms. The influence of nonradiative recombination and the bonding strength of atoms in an InGaN/GaN material system are lower and higher than those in AlGaAs/GaAs and InGaAsP/InP material system, respectively.

The defect motion and heat generation caused by nonradiative recombination also relates to the stability of the facet of laser diodes. The facet oxidation results from a chemical reaction enhanced optically. The ionization of host atoms is enhanced by the optical excitation of electrons from the valence to conduction band. The heating due to nonradiative recombination enhances the oxidation reaction at the facet. This nonradiative reaction tends to be intensified as the defective oxide film grows. In AlGaAs/GaAs material systems, the facet oxidation is easily generated because the material easily reacts with oxygen, and the high rate of nonradiative recombination occurs at the facet. The facet of InGaN/GaN laser diodes will be, therefore, stable when compared to that of AlGaAs/GaAs laser diodes.

Metals that are high in electronegativity, such as Au, easily react with semiconductors that have bandgap energies of approximately <2 eV even at room temperature (Hiraki *et al.* 1977; Hiraki 1980). An ohmic contact of alloy-type electrode is easily formed on the semiconductors having small bandgap energies such as GaAs and InGaAsP. This type of electrode is unstable under electrical bias and leads to device degradation. The alloy-type electrode is hardly formed on the semiconductors that have large bandgap energies, and the Schottky-type ohmic contact is usually formed on the semiconductors such as InGaN and GaN. The Schottky-type ohmic contact is stable under long-term operation.

The laser diodes composed of wide-gap semiconductors, therefore, tend to be stable during long-term operation.

1.3.3.2. *Sensing application*

In the fields of sensing applications or consumer electronics, the progress is significantly enhanced by employing the strained quantum-well structures. To extend the lasing wavelength band, relatively large mechanical stress is introduced to the active layer in the laser diodes. Lasing at around 2006 nm was obtained by introducing strain to the quantum well (active layer) in InGaAsP/InP laser diodes. The degradation behaviors of the strained quantum well laser diodes are similar to those of bulk-type laser diodes that are described above. Although these laser diodes operate stably, the wavelength scanning range for sensing gradually tends to narrow under degradation. This corresponds to the increase in the threshold current and the decrease in slope efficiency, as shown in Figure 1.19.

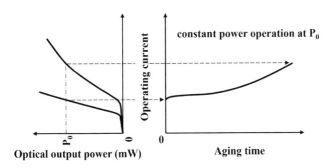

Figure 1.19. *Current-optical power relation and aging characteristics*

The characteristics required for sensing are lasing spectral linewidth, current-light output relationship, extended lasing wavelength range, etc. Gas-absorption spectra corresponding to the fundamental vibration are located in the infrared wavelength range and monitored by electrically and thermally scanning the wavelength. For example, the strong absorption band of CO_2 gas is located at around 2005 nm (see Figure 1.20). The sweep of the wavelength of the laser diode was performed by changing the temperature of the laser diode, and thus the light intensity gradually decreases as the wavelength lengthens. After degradation, the wavelength scanning range, which is changed with temperature, is gradually narrowed.

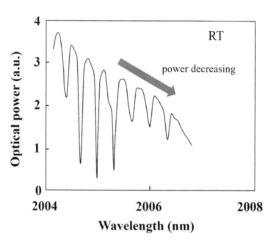

Figure 1.20. *CO_2-gas absorption spectrum monitored with a strained InGaAs laser diodes. The wavelength sweep was performed by changing the temperature of the laser diode at a constant injected current*

For sensing applications, various quantum cascade lasers lasing at the near- and middle-infrared wavelength bands have been developed and practically used in sensing equipment. These lasers stably operate because no carrier-injection and no recombination processes are employed for lasing.

To investigate the degradation behavior of the optical emitters, a failure analysis could be carried out. It is a scientific procedure able to localize the damage, recognize the link between the failure modes and the failure mechanism and propose corrective actions to avoid the problem in the future. Specific sample preparation and analysis, among the most relevant in the diagnostics of LDs, are proposed in the following in order to explain the physics of the failure responsible for several failed lasers.

1.4. Physical analysis of degraded lasers

Many techniques are considered suitable for investigating the failure modes and degradation mechanisms in lasers emitting diodes (LDs): electroluminescence (EL), photoluminescence (PL), optical beam induced current (OBIC), electron beam induced current (EBIC), cathodoluminescence (CL) and thermally induced voltage alteration (TIVA) (Ueda *et al.* 1985; Herrick *et al.* 1995; Magistrali *et al.* 1997; Mallard and Clayton 2004; Takeshita *et al.* 2010; Herrick 2011; Meneghini *et al.* 2013; Marioli *et al.* 2015; De Santi *et al.* 2016; Souto *et al.* 2016; Bushmaker *et al.* 2019).

Some of these inspection techniques require a special sample preparation that could be destructive, some enable information from the top, the back and from other sides. They could require the use of an optical microscope, a scanning electron microscope (SEM), a transmission electron microscope (TEM) or a focused ion beam (FIB). All of them are commonly used in failure analysis.

Among the most relevant in the diagnostics of LDs, the combination of focused ion beam for the specimen preparation and of the capability of scanning-transmission electron microscopy (S-TEM) to investigate specimens as thick as 1 µm have proved to be successful. Some excellent examples of this approach have been illustrated for the first time in Stark *et al.* (2005).

In the optoelectronics field, high spatial resolution instruments are fundamental to identify and analyze the root cause responsible for the observed failure.

Quite often the cause of failure in LDs is hidden below the surface. The SEM inspection is not able to reveal it and the TEM observation is generally limited by the spatial dimensions of a typical TEM lamella due to the undesired "potato-chipping" effects. However, the S-TEM microscopy, having minimal sensitivity to sample bending, enables a larger observation of a larger sample prepared as a TEM plan-view (PV).

It takes advantage of the capability of analyzing thick specimens prepared by an FIB, up to 1 µm, without the constraint of thin lamellas.

The possibility to observe in S-TEM mode lamellas wide and thick enough to include the whole active region of a laser diode in PV can provide, for example, important information on the propagation of the damages inside the active region. Examples will be discussed below. Besides, once the defect is entirely localized in a PV, a cross-section view or a longitudinal view could be taken in a specific area for further investigation.

Two failure mechanisms in LDs have specifically shown a wide extension that can be better analyzed by using the above-mentioned technique: catastrophic optical damages (CODs) and electrostatic discharges (ESDs). The direct observation of the defects structure could enable the visualization of complex liquid–solid combination to point out criticalities or to localize the formation of a latent defect that can cause the failure of the device in field operation.

Moreover, in the case of a COD event, the information obtained by taking both a PV and a cross-sectional view of the same defect has been decisive in the following examples to provide the final interpretation.

When applied to ESD damages, that sample preparation enabled a complete overview of the area of interest including the whole 3D extension of several defects. It was fundamental for the detection of some dangerous defects induced by a qualification test, not properly screened and consequently ready to reach the market, as the next example will show.

1.4.1. *Catastrophic optical damage*

COD generally indicates the instantaneous and irreversible deterioration in the LD efficiency that results when the power density at the laser facet or inside the cavity exceeds some critical value. The presence of a defect or inhomogeneity could lead to an increased local temperature. During a COD, the material is thermally damaged by the optical emission of the laser diode and critical melting temperature is reached.

The COD is considered one of the main limiting factors for reaching ultrahigh optical powers in LDs and it is identified as a major bottleneck for their reliability. For this reason, it has been diffusely studied over the last decades and several papers have been published on that subject (Cooper *et al.* 1966; Tiller 1968; Kressel *et al.* 1970; Shaw and Thornton 1970; Ettenberg *et al.* 1971; Hakki and Nash 1974; Henry *et al.* 1979; Ueda *et al.* 1985; Moser *et al.* 1991; Fukuda *et al.* 1994; Bou Sanayeh 2008; Bou Sanayeh *et al.* 2008; Ziegler *et al.* 2008a, 2008b; Tomm *et al.* 2009, 2011; Chin and Bertaska 2013; Hempel *et al.* 2013; Mura *et al.* 2013; Ueda and Pearton 2013). An accurate overview of the mechanism that takes place and its fast kinetics in GaAs-based LD is exemplarily summarized by Tomm *et al.* (2011).

The COD is an event described as the sudden drop-off of the optical power that takes place without warning and is generally characterized by three phases:

1) the facet temperature reaches a critical value and the defects in the material become absorbing site leading to an increase in the local temperature;

2) the thermal runaway process takes place on the facet or inside the bulk and a melted "hot spot" spreads spatially;

3) a further degradation continues in case the bias is not stopped immediately.

After a COD event, DLDs are produced during the laser operation; these DLDs are regions of the active zone of the laser without emission, they are locally generated, either at the front facet or inside the cavity, and can propagate along the cavity driven by the optical field.

A COD can occur at the laser mirror and the damage can further extend from the surface along the cavity. The energy for the defect generation/extension is provided by the laser emission. COD-related defects are nonradiative dislocations, which start from the output facet and propagate inside the cavity. Moreover, they can originate from some hot spots inside the bulk of the cavity, which are caused by highly nonradiative crystalline phase changes and structural defects such as dislocations.

An example of a COD-related damage in InP-based directly modulated-DFB (DM-DFB) edge-emitting laser diode has been proposed in Mura et al. (2013). A burnout occurred not at the mirror facet, but well inside the ridge, just at the edge of a thicker Au metallization covering the thinner metal (Figure 1.21). The longitudinal strain introduced by the thick metal edge was considered responsible for the distortion able to create a virtual partially reflecting mirror under the thick metal edge that makes the structure prone to parasitic light scattering.

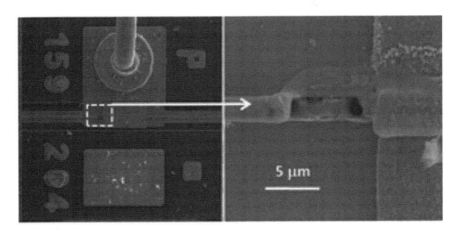

Figure 1.21. *SEM images of a COD on a DFB device occurring inside the ridge length, at the step of a thick metallization over the thinner one*

A COD is an event that can immediately destroy the cavity; at least three possible evolutions are then possible:

1) In case of a constant optical power operation mode, the reduced efficiency could induce an increase in the current leading the device to fail quickly.

2) In case of a transient overstress, the melting areas collapse one inside the other and a thermomechanical stress wave able to produce dislocations around the melting region could even determine the ejections of molten material out of the cavity.

In this context, a contribution (Mura *et al.* 2017) showed the damage pattern caused by the sudden degradation of the GaN-based high-power diode laser that was induced intentionally by a single-pulse step test scheme with a fixed operation current pulse length.

Figure 1.22 shows an SEM image of the front facet of the LD diode laser after sudden degradation together with the geometry of the device.

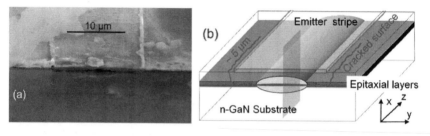

Figure 1.22. (a) COD pattern as observed by SEM at the front facet of a device. (b) Scheme illustrating the geometry of the device with the epitaxial layer on top, and the damage pattern. Cracks along the emitter stripes, including their extension along the z-axis, are marked in red. The reddish plane illustrates how the FIB trenches (x–z plane) that have been analyzed have been extracted. The coordinate system used is also given. For a color version of this figure, see www.iste.co.uk/vanzi/reliability.zip

The sample preparation by using an FIB has enabled a detailed cross-section inspection along with the defect that was induced by one singular COD event. After the cross-sectioning, SEM images are very effective at analyzing the laser degradation and together with Energy Dispersive Spectroscopy (EDS) X-ray spectroscopy provide an overview of the empty channel enabling the estimation in grams of the "missing" material (Figure 1.23).

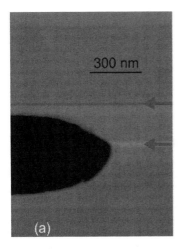 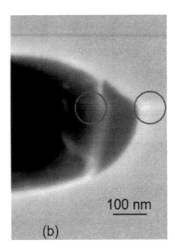

Figure 1.23. *On the left, a SEM image of FIB trench number 3 at the end of the channel that has been created by the COD process. Red arrows mark the electron blocking layer (dark) and the MQW (bright), the latter still not resolved. On the right, an S-TEM image of the FIB trench at the end of the channel. It can be observed that the MQW layers are clearly resolved. Both figures show the end of the channel exactly at the position of the MQW with a more pronounced widening toward the n (bottom) than toward the p-side (top). For a color version of this figure, see www.iste.co.uk/ vanzi/reliability.zip*

In Figure 1.23(b), a 200 KeV S-TEM analysis reveals the details of the epitaxial architecture as well as the position of the damage pattern.

The characterization of the event induced in a GaN LD is proposed in Tomm *et al.* (2018) and Tomm *et al.* (2017) as a comparison with previous studies carried on a GaAs LD under comparable experimental conditions. Even if COD in GaAs- and GaN-based diodes follow similar scenarios, the damage pattern differs substantially. The main differences can be summarized as follows:

i) the temperatures reached during COD in GaN are higher than these observed for GaAs-based devices;

ii) the damage patterns in GaN shows the formation of an empty channel (consistent with material loss due to ejections of hot material out of the front facet of the device) and no transition region/recrystallized material is observed. It differs from that observed in COD events in GaAs;

iii) average propagation velocities of defect fronts in degrading GaN-based devices are faster by a factor of 4–5 than what has been observed in GaAs-based devices.

3) The local melting destroys the resonance condition in the Fabry–Perot laser and modifies it in the DFB laser.

A melt/regrowth process takes place and is able to:

– reconstruct a solid crystal through the liquid phase epitaxy (LPE);

– blend all species (e.g. In, Ga, As, P);

– destroy any epitaxial stack and attempt lattice matching with the surrounding regular crystal.

In this condition, if the power (electrical or optical) is not switched off, an extended network of dislocations can further propagate. Moreover, during the solidification phase, the recrystallized material could even cause an "alternative" optical resonance condition inside the cavity.

An inspiring paper (Vanzi *et al.* 1991) described in detail the local melting phase, the crystal regrowth and the subsequent formation of defects in some degraded Buried Crescent (BC) InP/InGaAsP laser diodes.

In that work, the ellipsoidal hole caused by an optically induced damage was investigated in terms of its mathematical and physical evolution and the expected defect distribution in InP- and GaAs-based devices. The identification of the melt/regrowth mechanism is crucial for attributing the root cause of the failure to the optical power.

The fingerprint is the ellipsoidal region where epitaxy is melted down. The authors demonstrated that this interpretation is more than qualitative. If a melt/regrowth process takes place, the regrowth of the mixture of In, Ga, As and P will segregate the low-melting indium to the ultimate steps. This process, in turn, because of the bigger size of the atoms, will cause dense defects along the ultimate solidification line.

These results have been recalled in a recent paper on COD in InP-based emitters (Vanzi *et al.* 2016b) where three types of field failures return due to COD events. They have been proposed and explained below.

1.4.1.1. *First case*

Some InP-based 1310 nm DFB emitters failed in field operation. They showed a second peak in the spectrum (Figure 1.24), far from the surviving original one (at a distance of 1 nm). It is known that in a DFB, the feedback is distributed along the whole cavity length, so it is possible that the devices, after the damage, continue displaying a measurable threshold and some residual laser action.

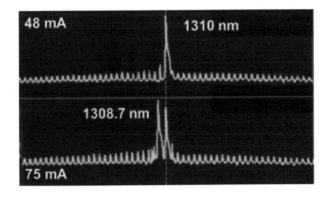

Figure 1.24. *Spectrum at two different driving currents of the same single mode laser*

This bimodality in a single-mode device is easily explained observing the structure of the ridge in PV (Figure 1.25). The observation of the entire area where the COD takes place helps to identify the peculiar regrowth kinetics after the initial fusion.

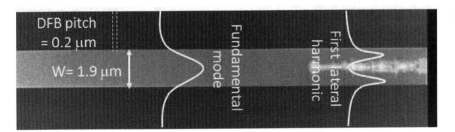

Figure 1.25. *S-TEM plan view on the rear facet of a failed LD, showing a typical COD propagating from the right side boundary of the optical cavity to inside*

In Figure 1.26, the damage is proposed in PV and longitudinal cross-section (LX) view, and the samples were taken from two different devices. Besides, in Figure 1.27, the enlarged view of the previous LX image clearly shows the alternate regions where the multi quantum well (MQW) stack survives or disappears, which are distributed with the same periodicity of the DFB grating.

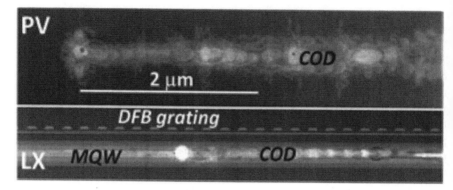

Figure 1.26. *PV and LX S-TEM on the rear facets of two failed devices*

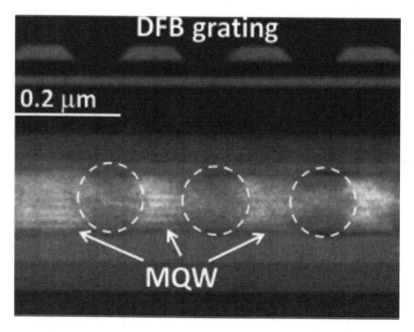

Figure 1.27. *Details of the LX view proposed in Figure 1.6. The alternation of presence and disappearance (circled areas) of MQWs is clear*

The disappearance of the MQW and the absence of any extended lattice defect indicate that the melting/regrowth process was driven by the optical field as a thermal source.

As soon as the material reaches its melting point, the light generation locally and suddenly switches off, and the fused region is cooled down.

It is worth noticing that the materials that compose the active region and the confinement layers are now blended in a liquid phase and are confined inside several melting pots. These are shaped as the isothermal surfaces generated by the optical field and whose boundaries preserve the lattice of the solid structure. This means that the intermixed materials, cooling down, solidify under LPE conditions. The process starts from the liquid–solid interfaces to inside, following some complicated phase diagrams that will segregate the lowest melting material from the center of the regrown region.

Due to this peculiar damage pattern, the interpretation of the observed spectrum is the following: at low injection current, the light oscillation inside the damaged segment of the active cavity is mostly switched off, and the surviving cavity on the left sustains the regular oscillation at the wavelength fixed by the DFB corrugation. At higher injection conditions, the fundamental transversal mode is attenuated but still active. In this condition, the first lateral mode can take place and its maxima are symmetric to the axis.

The quantitative shift of the second peak proved to be consistent with the excitation of the second lateral mode of a Hermite–Gaussian beam for the specific material and geometries.

In conclusion, the reason for that transition was the partial damage of the cavity. It caused the partial suppression of the fundamental transversal mode and allowed the hopping to the second mode at high injection. The current example is easily explained as a COD-induced higher order lateral mode.

1.4.1.2. Second case

Some damages occurred in a set of 980 nm pump LD, externally tuned using a fiber DBR.

The analysis showed that the reason for the failure was that a COD was able to induce a vertical higher order mode that sequentially causes a second-level COD.

The damage observed in Figure 1.28 suggests that the first event was a standard COD because it caused the ellipse aligned with the active layer. To explain the damage below the active layer, it is necessary to suppose that the light did not disappear in this area, but that an intense optical beam traveled below the molten/regrown region, causing a second COD. This second COD will melt the second elliptic region, and the upper side of the new liquid/solid interface runs along the defect line of the first ellipse is the key point. The evolving LPE process does not replicate a perfect lattice but will cause a complex network of twins and thread dislocations to fill the space between the major axes of the two ellipses.

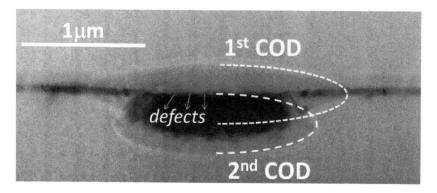

Figure 1.28. *COD two-step kinetics*

The conclusion is that CODs can induce modifications of the optical cavity, which is able to excite higher harmonics. This is quickly observed in Figure 1.28 which shows the regrowth under the LPE process where the molten/regrown region is the result of the effect of the two CODs.

1.4.1.3. *Third case*

A high-power 1310 nm DFB structure failed during a high-temperature operating life (HTOL) test. S-TEM PV in Figure 1.29 shows a long COD characterized by two parallel rails, which at certain point converge into a single rail, occupying half the width of the ridge-guided area.

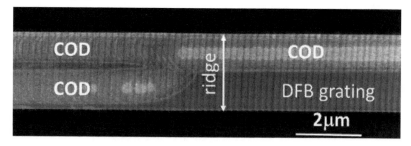

Figure 1.29. *S-TEM PV of a DFB structure showing a COD separated into two parallel paths*

It is worth noting that the sample preparation can reveal the entire area of the damage, and some cross-sectional views are possible at the points of interests.

Figure 1.30 provides X-TEM views of double and single rails.

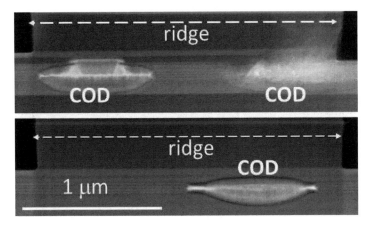

Figure 1.30. *X-TEM views of double (above) and single rails (below)*

No primitive damage has been found on the failed devices, but only the observed long damaged paths, that in turn cannot be anything other than molten/regrown regions following the optical intensity.

In that case, the interpretation is that the COD is not the cause, but the effect of the higher order transversal optical mode. This case must be considered as a lateral mode induced COD. It witnesses the firing of a dominating transversal second harmonic as a precursor of the catastrophic event.

Moreover, the observed extension of these defects suggests that the presence of local native defects inside the cavity is not mandatory because the power density itself can provide the energy necessary to melt the optical peak areas. However, any design solution or process flaw able to break the axial symmetry of the optical cavity could be the root cause of exciting the side modes.

In the proposed cases, the high power is responsible for the COD event. In the first example, even if the LDs were operating at low power, the presence of a second peak reveals the higher order mode that can appear only when the driving current increases.

The conclusion is that the high optical power density is the trigger for higher harmonics, which not only includes high power devices, but also the low power ones whose active region is highly confined both vertically and laterally.

The final result that the authors propose to the reader is that a new failure mechanism should be added to the list of potential physical causes of COD events.

The higher order transversal optical mode is certainly the cause responsible for the former case proposed above.

Proper operation conditions can reduce the chance to cause a COD, and important technological improvement has succeeded in improving the robustness of LDs against COD. Among those we can quote different facet passivation techniques and non-absorbing mirror schemes, as well as appropriate reflectivity coating configurations and minimization during crystal growth of the generation of defects participating in nonradiative recombination (Holonyak 1985; Glasser and Latta 1991; Epperlein 2013).

Even if these improvements and studies have been successfully verified through novel devices that have significantly higher brightness, up to now the general opinion is still that the COD is an effect that cannot be completely eliminated, but only delayed toward higher photon densities.

The reason why we are still dealing with COD is the need to better understand the causes behind that failure mode and the need for strategies able to decrease the occurrence of this catastrophic problem and to extract more power from LDs. The challenge in improving the device performance remains the prevention of the initial thermal runaway.

In the abovementioned case studies, the sample preparation together with the consequently microscopic analysis by using S-TEM has proven to be fundamental to understand how the CODs took place.

1.4.2. *Electrostatic discharge*

The ESD damage is defined as "the change to an item caused by an ESD that makes it fail to meet one or more specified parameters" and can occur at any point from manufacturing step to field service (ESD Association 2013). An accurate description of an ESD, its physics and the fundamentals are proposed in Voldman (2004) and Amerasekera *et al.* (1992).

ESD damage is usually caused by one of three events: direct ESD from human to the device, ESD from the charged device or field-induced discharges. For that reason, the suitable models to simulate ESD-induced damages are called the human body model (HBM), the charged device model (CDM) and the charged machine model (MM).

ESD damage is recognized as a major source affecting the lifetime of the oxide in vertical cavity surface emitting lasers (VCSELs).

VCSELs are rapidly diffusing in several fields, including remote sensing, illumination, facial recognition technology, data storage and short-haul high data rate fiber communications, and because of some distinctive features offered by VCSEL technology, such as high speed, circular beam shape, low cost, and small size, they cover a range of relevant mass applications, from datacom to sensing in the last generation of smartphones.

In case of an ESD event, either catastrophic failures or latent defects are possible. While the first ones are easy to detect, the former ones are extremely difficult to prove. Studies on technological improvements in ESD robustness of laser diodes have been available in the literature for a long time, and the results of some tests are used to verify product reliability compliance with the Bellcore or MIL-STD-883 standards (MIL STD 1989; EIA/JESSD22-A114 1996; ESD Association 1996a, 1996b, 1998; IEC 2008).

A recent paper (Vanzi et al. 2016a) proposed the analysis of several failures detected after the qualification plan for the telecom application of some commercial GaAs-based VCSELs emitting at 850 nm.

As mentioned above, VCSELs diodes have a reputation for being very sensitive to ESD (Neitzert et al. 2001; Krueger et al. 2003; McHugo et al. 2003).

The qualification tests included several transient electrical stress tests: forward and reverse ESD tests according to the HBM, the CDM and MM, surge test and over power (OVP) test.

The summary proposed here aims to point out that the interpretation of the physical mechanism responsible for the observed damages has been possible because of an accurate FIB sample preparation and an S-TEM analysis. The confirmation was then provided by a device modeling.

After the tests, an electrooptical characterization and a physical analysis was performed. The investigation performed on the devices is illustrated in the following lines together with the most evident and contradictory results.

The failure criterion was a certain variation of the optical power (in terms of threshold current and optical efficiency) and a leakage level larger than 2 µA at −5 V.

The qualification tests caused several different patterns of physical damage and different degrees of electrooptical degradation.

Some damage patterns, well-known in the VCSEL literature (Cheng *et al.* 1996; Hawkins *et al.* 2002; Herrick 2002; Mathes *et al.* 2005; Kim *et al.* 2006, 2008; Ueda and Pearton 2013), appeared in devices with extremely different leakage levels, and surprisingly even in some devices that passed the stress test.

These devices passed the ESD test because, in terms of emitted optical power and current leakage, they did not show any change. The possibility that the tests introduced some undetected damages was considered.

The physical analysis was driven by using the forward EL inspection and the reverse-bias emission microscopy. Subsequently, the FIB and 200 KeV S-TEM in z-contrast mode analysis was performed to inspect the pattern of the observed damages.

As an example, Figure 1.31 reports two kinds of physical damage, corresponding to 200 V (case A) and 1000 V (case B) reverse HBM ESD test: EL analysis on the left, and S-TEM analysis on a PV thick lamella on the right.

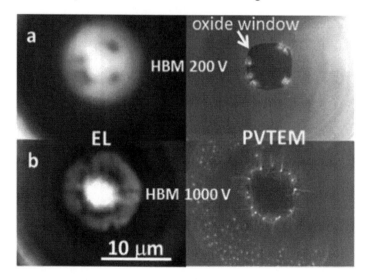

Figure 1.31. *Electroluminescence and planar view S-TEM on two devices after reverse HBM test at (a) 200 V and (b) 1000 V*

Despite the obvious damages around the emitting area, case A did not show any significant change in the optical characteristics except a small leakage of <5 µA at −5 V.

Case B, however, displayed very large damage confirmed by the total loss of the optical emission and leakage at −5 V as large as about 240 µA.

Case A passed the test and B failed it. It is worth noting the apparent poor correlation between the physical damage and the optical degradation in case A. It leads to the conclusion that the higher stress (1000 V) created the damage outside the oxide window and it was responsible for the main electrical and optical degradation.

Besides, a further device stressed at 1000 V reverse HBM test did not show any change after the test, but, after some ageing, it showed the same failure modes of the previous device. A combination of variety between the applied stresses, the physical evidence and the detectable optical and electrical failure modes was then observed.

A further cross-section was performed. Figure 1.32 shows that the dark spots that appear along the internal rim of the oxide windows in Figure 1.31(a) are vertical structures surrounded by nested systems of dislocations.

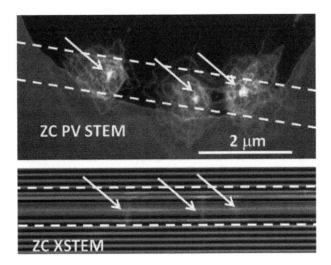

Figure 1.32. *Plan view (PV) and X-TEM view at S-TEM in Z-contrast (ZC) of two similar devices that show three similar ESD-induced defects (arrowed in the pictures). The dashed lines in each image include the whole 3D extension of the defects*

Due to their position at the boundary of the optical cavity, they are not able to cause any relevant electrical or optical effect. Nevertheless, they could easily become nucleation points for growing defects in field operation.

The damages observed in Figure 1.31(b), outside the oxide window, may appear at a different extent, but always involve the perforation of the oxide. It is observed that for the weaker cases, the perforation creates small molten regions that, cooling down, are likely to preserve the dominant local doping, and then re-build a parasitic pn-junction that keeps the measured leakage limited (Figure 1.33). Of course, high levels of stress can lead directly to a faulty state.

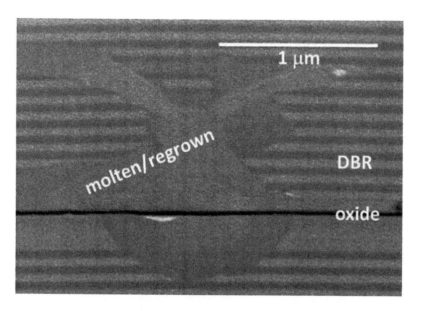

Figure 1.33. *Transversal view of an oxide damage, with the evidence of melting and growth of the layers of the DFB*

It should be recalled that some defects may occur without any external manifestation. Also, the types of transient stress may be different, leading to quite similar degradation states.

A circuit simulation was able to individuate four regions A, B, C (oxidized region) and D (active region) obtained by the increasing voltage across the nodes typical of an ESD discharge of a resistive network proposed for the VCSEL. That approach has proven to be effective in mapping the damages that could be induced because of an ESD event (see Figure 1.34).

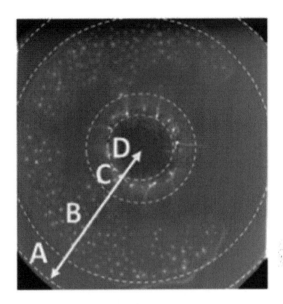

Figure 1.34. *Four regions A, B, C and D correspond to different patterns on a case reported in Figure 1.31*

These damages appear as the result of current filaments, causing a small damaged core and a surrounding region that is mostly made of nested dislocations for the active region and extended fusion/solidification areas in the oxidized areas.

Besides, the results point out that the weakest structure is the edge of the active region. In summary, failures distributed across the oxide always cause an excess of leakage and a reduced light emission. However, failures near the active region can occur as silent damages, not detected by the screening methods indicated by the international standards.

From a reliability point of view, their risk is higher when they grow and quickly propagate across the device in-field operation. It can kill it in a very short time.

The effectiveness of EL for the detection of silent damages is proved, although EL is normally not included in the standard procedures for the qualification. This study suggests its introduction as a screening procedure whenever possible, in particular, if the ESD risk is high.

Moreover, this work is a further confirmation of the peculiar features of ESD events in VCSELs.

It is worth noting that while most semiconductor lasers are quite vulnerable to ESD, it is still overdiagnosed by failure analysis (Mura and Vanzi 2007; Herrick 2011). ESD is the most frequent suspect of failures when the starting point of the dislocation is in a wrong place or its nucleus does not correspond to any of the structures shown in Figure 1.31. Some caution is required before holding ESD responsible for failure even when the damage peculiarities hardly justify such interpretation. Moreover, the study points out a severe warning for possible ESD-induced latent damages after an ESD test.

The possibility to observe in S-TEM mode the whole active region of a laser diode in PV has proved to be successful in the understanding of the propagation of the damages and the detection of latent defects. Further information can also be obtained by cross-sectioning the PV to isolate specific areas of interest.

1.5. References

Amerasekera, A., van den Abeelen, W., van Roozendaal, L., Hannemann, M., and Schofield, P. (1992). ESD failure modes: Characteristics mechanisms, and process influences. *IEEE Trans. On Electron Dev.*, 39(2), 430–436.

Ando, K., Yamaguchi, M., and Uemura, C. (1984). Impurity (Si) concentration effects on radiation-induced deep traps in n-type InP. *J. Appl. Phys.*, 55(12), 4444–4446.

Asada, M. and Suematsu, Y. (1983). The effect of loss and nonradiative recombination on the temperature dependence of threshold current in 1.5-1.6 µm GaInAsP/InP lasers. *IEEE J. Quantum Electron.*, QE-19, 917–923.

Bou Sanayeh, M. (2008). Catastrophic optical damage in high-power, broad-area laser diodes. PhD Dissertation, Von der Fakultät für Ingenieurwissenschaften der Universität Duisburg-Essen.

Bou Sanayeh, M., Brick, P., Schmid, W., Mayer, B., Müller, M., Reufer, M., Streubel, K., Ziegler, M., Tomm, J.W., and Bacher, G. (2008). The physics of catastrophic optical damage in high-power AlGaInP laser diodes. *Proc. SPIE 6997, Semiconductor Lasers and Laser Dynamics III*, 699703.

Bushmaker, A.W., Lingley, Z., Brodie, M., Foran, B., and Sin, Y. (2019). Optical beam induced current and time resolved electro-luminescence in vertical cavity surface emitting lasers during accelerated aging. *IEEE Photonics J.*, 11(1.5) 1–12.

Cheng, Y.M., Herrick, R.W., Petrof, P.M., Hibbs-Brenner, M.K., and Morgan, R.A. (1996). Degradation mechanisms of vertical cavity surface emitting lasers. *IEEE International 34th Annual Proceedings on Reliability Physics Symposium*, Dallas, Texas, 211–213.

Chin, A.K. and Bertaska, R.K. (2013). Catastrophic optical damage in high-power, broad-area laser diodes. In *Materials and Reliability Handbook for Semiconductor Optical and Electron Devices*, Ueda, O. and Pearton, S.J. (eds). Springer, New York, 123–145.

Cooper, D.P., Gooch, C.H., and Sherwell, R.J. (1966). Internal self-damage of gallium arsenide lasers. *IEEE J. Quantum Elect.*, QE-2(8), 329–330.

De Santi, C., Meneghini, M., Gachet, D., Mura, G., Vanzi, M., Meneghesso, G., and Zanoni, E. (2016). Nanoscale investigation of degradation and wavelength fluctuations in InGaN-based green laser diode. *IEEE Trans. Nanotechnol.*, 15(2) , 274–280.

EIA/JESSD22-A114 (1996). JESD22-A114E. Datasheet, January 2007.

Epperlein, P.W. (2013). Semiconductor laser engineering, reliability and diagnostics: A practical approach to high power and single mode devices. DOI: 10.1002/ 9781118481882, February 2013.

ESD Association (1996a). ESD-DS5.2. Sensitivity Testing, Machine Model. ESD Association, Rome.

ESD Association (1996b). ESD DS5.3.1. Device Testing: Charged Device Model. ESD Association, Rome.

ESD Association (1998). ESD-STM5.1. Sensitivity Testing, Human Body Model (HBM). ESD Association, Rome.

ESD Association (2013). Fundamentals of Electrostatic Discharge Part One – An Introduction to ESD© 2013. ESD Association, Rome.

Ettenberg, M., Sommers, H.S., Kressel, H., and Lockwood, H.F. (1971). Control of facet damage in GaAs laser diodes. *Appl. Phys. Lett.*, 18, 571–573.

Ettenberg, M., Nuese, C.J., and Kressel, H. (1979). The temperature dependence of threshold current for double heterojunction lasers. *J. Appl. Phys.*, 50, 2949–2954.

Fukuda, M. (1991). *Reliability and Degradation of Semiconductor Lasers and LEDs*. Artech House, Boston/London.

Fukuda, M. (2000). Historical overview and future of optoelectronics reliability for optical fiber communication systems. *Microel. Reliab.*, 40, 27–35.

Fukuda, M. (2007). Reliability of semiconductor lasers used in current communication systems and sensing equipment. *Microel. Reliab.*, 47, 1619–1624.

Fukuda, M. (2009). Reliability issues of light emitting, amplifying, and modulating semiconductor devices. *Proceedings of the International Symposium on Reliability of Optoelectronics for Space (ISROS 2009)*, Sardinia, Italy, 46–55.

Fukuda, M., Hirono, T., Kurosaki, T., and Kano, F. (1993). 1/f noise behavior in semiconductor laser degradation. *IEEE Photon. Tech. Lett.*, 5, 1165–1167.

Fukuda, M., Okayasu, M., Temmyo, J., and Nakano, J. (1994). Degradation behavior of 0.98-μm strained quantum well InGaAs/AlGaAs lasers under high-power operation. *IEEE J. Quantum Electronics*, 30, 471–476.

Fukuda, M., Mishima, T., Nakayama, N., and Masuda, T. (2010). Temperature and current coefficients of lasing wavelength in tunable diode laser spectroscopy. *Appl. Phys. B*, 100, 377–382.

Glasser, M. and Latta, E.E. (1991). Method for mirror passivation of semiconductor laser diodes. U.S. Patent 5,063,173.

Goodwin, A.R., Peters, J.R., Pion, M., Thompson, G.H.B. (1975). Threshold temperature characteristics of double heterostructure GaAlAs lasers. *J. Appl. Phys.*, 46, 3126–3130.

Hakki, B.W. and Nash, F.R. (1974). Catastrophic failure in GaAs double-heterostructure injection lasers. *J. Appl. Phys.*, 45(9), 3907–3912.

Hawkins, B.M., Hawthorne, R.A., Guenter, J.K., Tatum, J.A., and Biard, J.R. (2002). Reliability of various size oxide aperture VCSELs. *52nd Electronic Components and Technology IEEE Proceedings*, 540–550.

Hempel, M., Tomm, J.W., La Mattina, F., Ratschinski, I. and Schade, M. (2013). Microscopic origins of catastrophic optical damage in diode lasers. *J. Sel. Top. Quant. Electron.*, 19(4), 1500508.

Henry, C.H., Petroff, P.M., Logan, R.A., and Meritt, F.R. (1979). Catastrophic damage of $Al_xGa_{1-x}As$ double-heterostructure laser material. *J. Appl. Phys.*, 50, 3721–3723.

Herrick, R.W. (2002). Oxide VCSEL reliability qualification at Agilent Technologies. *Proceedings of SPIE 4649, Vertical-Cavity Surface-Emitting Lasers VI.* DOI: 10.1117/12.469227.

Herrick, R.W. (2011). Failure analysis and reliability of optoelectronics devices. In *Microelectronics: Failure Analysis Desk Reference*, 5th edition EDFAS Desk Reference Committee, ASM International.

Herrick, R.W., Cheng, Y.M., Petroff, P.M., Hibbs-Brenner, M.K., and Morgan, R.A. (1995). Spectrally-filtered electroluminescence of vertical-cavity surface-emitting lasers. *IEEE Photonics Tech. Lett.*, 7(110), 1107–1109.

Hiraki, A. (1980). A model on the mechanism of room temperature interfacial intermixing reaction in various metal-semiconductor couples: What triggers the reaction? *J. Electrochem. Soc.*, 127, 2662–2665.

Hiraki, A., Shuto, K., Kim, S., Kammura, W., and Iwami, M. (1977). Room temperature interfacial reaction in Au-semiconductor system. *Appl. Phys. Lett.*, 31, 611–612.

Holonyak, N. Jr. (1985). Semiconductor device fabrication with disordering elements introduced into active region. U.S. Patent 4,511,408.

IEC (2008). IEC 61000-4-2: IEC 61000-4-2 (2.0 ed.). International Electrotechnical Commission (IEC). December 2008, 7, 10–13.

Ikoma, N., Kawahara, T., Kaida, N., Murata, M., Moto, A., and Nakabayashi, T. (2005). Highly reliable AlGaInAs buried heterostructure lasers for uncooled 10 Bb/s direct modulation. *Technical Digest of OSA/IEEE Conference on Optical Fiber Communication*, Paper OThU1, Anaheim, CA.

Johnston, A.H. and Miyahira, T.F. (2000). Characterization of proton damage in light-emitting diodes. *IEEE Trans. Nucl. Sci.*, 47(6), 2500–2507.

Johnston, A.H. and Miyahira, T.F. (2004). Radiation degradation mechanisms in laser diodes. *IEEE Trans. Nucl. Sci.*, 51(6), 3564–3571.

Johnston, A.H., Miyahira, T.F., and Rax, B.G. (2001). Proton damage in advanced laser diodes. *IEEE Trans. Nucl. Sci.*, 48(6), 1784–1772.

Khanne, R., Allums, K.K., Abernathy, C.R., and Pearton, S.J. (2004). Effects of high-dose 40 MeV proton irradiation on the electroluminescent and electrical performance of InGaN light-emitting diode. *Appl. Phys. Lett.*, 85(15), 3131–3133.

Kim, H.D., Jeong, W.G., Shin, H.E., Ser, J.H., Shin, H.K., and Ju, Y.G. (2006). Reliability in the oxide vertical-cavity surface-emitting lasers exposed to electrostatic discharge. *Opt. Express*, 14(25), 12432–12438.

Kim, T., Kim, T., Kim, S., and Kim, S.B. (2008). Degradation behavior of 850 nm AlGaAs/GaAs oxide VCSELs suffered from electrostatic discharge. *ETRI J.*, 30(6), 833–843.

Kressel, H., Nelson, H., and Hawrylo, F.Z. (1970). Control of optical losses in p-n junction lasers by use of a heterojunction: Theory and experiment. *J. Appl. Phys.*, 41, 2019–2031.

Krueger, J., Sabharwahl, R., McHugo, S.A., Tan, N., Janda, N., Mayonte, M.S., Heidecker, M., Eastley, D., Keever, M.R., and Kocot, C.P. (2003). Studies of ESD-related failure patterns of Agilent oxide VCSELs. *SPIE Proc. 4994*, 162–172.

Le Metayer, P., Gilard, O., Germanicus, R., Campillo, D., Ledu, F., Cazes, J., Falo, W., and Chatry, C. (2003). Proton damage effects on GaAs/GaAlAs vertical cavity surface emitting lasers. *J. Appl. Phys.*, 94(12), 7757.

Lang, D.V. and Kimerling, L.C. (1974). Observation of recombination-enhanced defect reactions in semiconductors. *Phys. Rev. Lett.*, 33(8), 489–492.

Lippen, T.V. (2002). Electrical isolation of AlGaAs by ion irradiation. *Appl. Phys. Lett.*, 80(2), 264–266.

Maeda, K. and Takeuchi, S. (1996). Enhancement of dislocations mobility in semiconducting crystal by electronic excitation. In *Dislocation in Solids*, Nabaro, F.R.N. and Duesbery, M.S. (eds). North-Holland, Amsterdam, 435–504.

Magistrali, F., Salmini, G., Martines, G., and Vanzi, M. (1997). Charge diffusion and reciprocity theorems: A direct approach to EBIC of ridge laser diodes. *Proc of ISTFA 97*, 233–238.

Mallard, R.E., and Clayton, R.D. (2004). EBIC and TEM analysis of catastrophic optical damage in high-power GaAlAs/GaInAs lasers. *Proc. SPIE 3004, Fabrication, Testing, and Reliability of Semiconductor Lasers II*, 1997. DOI: 10.1117/12.273827.

Marioli, M., Meneghini, M., Rossi, F., Salviati, G., de Santi, C., Mura, G., Meneghesso, G., and Zanoni, E. (2015). Degradation mechanisms and lifetime of state-of-the-art green laser diodes. *Phys. Status Solidi A., and Materials Science.* DOI: 10.1002/pssa.201431714.

Mathes, D., Guenter, J., Hawkins, B., Hawthorne, B., and Johnson, C. (2005). An atlas of ESD failure signatures in vertical cavity surface emitting lasers. *Proceedings of the 31st International Symposium for Testing and Failure Analysis.* ASM International, USA.

Mawatari, H., Fukuda, M., and Tohmori, Y. (1999). Degradation behavior of the active region and passive region in buried heterostructure (BH) distributed Bragg reflector (DBR) lasers. *Microel. Reliab.,* 39, 1857–1861.

McHugo, S.A., Krishnan, A., Krueger, J., Luo, Y., Tan, N-X., Osentowski, T., Xie, S., Mayonte, M., Herrick, B., Deng, Q., Heidecker, M., Eastley, D., Keever, M., and Kocot, C.P. (2003). Characterization of failure mechanisms for oxide VCSELs. *SPIE Proc. 4994,* 55–66.

Meneghini, M., Carraro, S., Meneghesso, G., Trivellin, N., Mura, G., Rossi, F., Salviati, G., Holc, K., Weig, T., Schade, L., Karunakaran, M.A., Wagner, J., Schwarz, U.T., and Zanoni, E. (2013). Degradation of InGaN/GaN laser diodes investigated by micro-cathodoluminescence and micro-photoluminescence. *Appl. Phys. Lett. 103,* 233506. DOI: 10.1063/1.4834697.

MIL STD 883.C/3015.7 notice 8 (1989). Military Standard for Test Methods and Procedures for Microelectronics: ESD Sensitivity Classification, U.S. Department of Defence.

Moser, A., Oosenbrug, A., Latia, E.E., Forster, T., and Gasser, M. (1991). High-power operation of strained InGaAs/AIGaAs single quantum well lasers. *Appl. Phys. Lett.,* 59, 2642–2644.

Mura, G. and Vanzi, M. (2007). Failure analysis of failure analyses: The rules of the Rue Morgue, ten years later. *IEEE Trans. Dev. & Mat. Reliab.,* 7, 446–452.

Mura, G., Vanzi, M., Marcello, G., and Cao, R. (2013). The role of the optical trans-characteristics in laser diode analysis. *Microel. Reliab,* 53, 1538–1542.

Mura, G., Vanzi, M., Hempel, M., and Tomm, J.W. (2017). Analysis of GaN based high-power diode lasers after singular degradation events. *Phys. Status Solidi RRL,* 11(7), 1700132.

Nakahara, K., Tsuchiya, T., Kitatani, T., Shinoda, K., Kikawa, T., Hamano, F., Fujisaki, S., Taniguchi, T., Nomoto, E., Sawada, M., and Yuasa, T. (2004). 12.5-Gb/s direct modulation up to 115C in 1.3-um InGaAlAs-MQW RWG DFB lasers with notch-free grating structure. *J. Lightwave Technol.,* 22(1), 159–165.

Neitzert, H., Piccirillo, A., and Gobbi, B. (2001). Sensitivity of proton implanted VCSELs to electrostatic discharge pulses. *IEEE J. Sel. Topics Quantum Electron,* 7, 231–241.

Okada, H., Nakanishi, Y., Wakahara, A., Yoshida, A., and Ohshima, T. (2008). 380 keV proton irradiation effects on photoluminescence of Eu-doped GaN. *Nuclear Inst. Methods Phys. Res.*, B266, 853–856.

O'Neill, J.P., Ross, I.M., Cullis, A.G. Wang, T., and Parbrook, P.J. (2003). Electron-beam-induced segregation in InGaN/GaN multiple-quantum wells. *Appl. Phys. Lett.*, 83(10), 1965–1967.

Petroff, P. and Hartman, R.L. (1973). Defect structure introduced during operation of heterojunction GaAs lasers. *Appl. Phys. Lett.*, 23(8), 469–471.

Petroff, P. and Hartman, R.L. (1974). Rapid degradation phenomenon in heterojunction GaAlAs-GaAs lasers. *J. Appl. Phys.*, 45(9), 3899–3903, 1974.

Petroff, P.M., Logan, R.A., and Savage, A. (1980). Nonradiative recombination at dislocations in III-V compound semiconductors. *Phys. Rev. Lett.*, 44(4), 287–291.

Schone, H., Carson, R.F., Paxton, A.H., and Taylor, E.W. (1997). AlGaAs vertical-cavity surface-emitting laser response to 4.5-MeV proton irradiation. *IEEE Photon. Technol. Lett.*, 9(12), 1552–1554.

Shaw, D.A. and Thornton, P.R. (1970). Catastrophic degradation in GaAs laser diodes. *Solid State Electron.*, 13(7), 919–922.

Souto, J., Pura, J.L., Torres, A., Jiménez, J., Bettiati, M., and Laruelle, F.J. (2016). Sequential description of the catastrophic optical damage of high power laser diodes. *Proc. SPIE 9733, High-Power Diode Laser Technology and Applications XIV*. DOI: 10.1117/12.2212953.

Stark, T.J., Russell, P.E., and Nevers, C. (2005). 3-D defect characterization using plan view and cross-sectional TEM/STEM analysis. *ISTFA Conference Proc. of ISTFA05*, San Jose, California.

Sugimura, A. (1981). Band-to-band Auger recombination effect on InGaAsP laser threshold. *IEEE J. Quantum. Electron.*, QE-17, 627–635.

Takeshita, T., Sato, T., Mitsuhara, M., Kondo, Y., Oohashi, H. (2010). Degradation analysis of InP buried heterostructure layers in lasers using optical-beam-induced-current technique. *IEEE Trans. Dev. Mat. Reliab.*, 10(1), 142–148.

Tan, H.H. and Jagadish, C. (1997). Wavelength shifting in GaAs quantum well lasers by proton irradiation. *Appl. Phys. Lett.*, 71(18), 2680–2682.

Tan, H.H., Williams, J.S., and Jagadish, C. (1996). Large energy shifts in GaAs-AlGaAs quantum wells by proton irradiation-induced intermixing. *Appl. Phys. Lett.*, 68(17), 2401–2403.

Tiller, W.A. (1968). Theoretical analysis of requirements for crystal growth from solution. *J. Cryst. Growth*, 2(2), 69–79.

Thompson, G.H.B. and Henshall, G.D. (1980). Nonradiative carrier loss and temperature sensitivity of threshold current in 1.27 μm (GaIn)(AsP) D. H. lasers. *Electron. Lett.*, 16, 42–43.

Tomiya, S., Hino, T., Goto, S., Yakeya, M., and Ikeda, M. (2004). Dislocation related issues in the degradation of GaN-based laser diodes. *J. Sel. Top. Quant. Electron.*, 10, 1277–1286.

Tomm, J.W., Ziegler, M., Talalaev, V., Matthiesen, C., Elsaesser, T., Bou Sanayeh, M., Brick, P., and Reufer, M. (2009). New approaches towards the understanding of the catastrophic optical damage process in in-plane diode lasers. *SPIE 7230, Novel In-Plane Semiconductor Lasers VIII*, 72300V.

Tomm, J.W., Ziegler, M., Hempel, M., and Elsaesser, T. (2011). Mechanisms and fast kinetics of the catastrophic optical damage (COD) in GaAs-based diode lasers. *Laser Photonics Rev.*, 5(3), 422–441.

Tomm, J.W., Kernke, R., Mura, G., Vanzi, M., and Hempel, M. (2017). Comparison of catastrophic optical damage events in GaAs- and GaN-based diode lasers. *IEEE High Power Diode Lasers and Systems Conference (HPD)*, Coventry, UK. DOI: 10.1109/HPD. 2017.8261097.

Tomm, J.W., Kernke, R., Mura, G., Vanzi, M., Hempel, M., Acklin, B. (2018). Catastrophic optical damage of GaN-based diode lasers: Sequence of events, damage pattern, and comparison with GaAs-based devices. *J Elec. Mater.*, 47. 4959. DOI: 10.1007/s11664-018-6144-6.

Ueda, O. and Pearton S.J. (2013). *Materials and Reliability Handbook for Semiconductor Optical and Electron Devices*. Springer Science & Business Media, New York.

Ueda, O., Wakao, K., Komiya, S., Yamaguchi, A., Isozumi, S., and Umebu, I. (1985). Catastrophic degradation of InGaAsP/InGaP double-heterostructure lasers grown on (001) GaAs substrates by liquid-phase epitaxy. *J. Appl. Phys.*, 58(11), 3996–4002.

Vanzi, M., Giansante, M., Zazzetti, L., Magistrali, F., Sala, D., Salmini, G., and Fantini, F. (1991). Interpretation of failure analysis results on ESD-damaged InP laser diodes. *Proc. of ISTFA91*, 305–314. ASM International, USA.

Vanzi, M., Mura, G., Marcello, G., and Xiao, K. (2016a). ESD tests on 850 nm GaAs-based VCSELs. *Microelectronics Reliability*, 64, 617–622.

Vanzi, M., Xiao, K., Marcello, G., and Mura, G. (2016b). Side mode excitation in single-mode laser diodes. *IEEE Trans. Device Materials Reliability*, 16(2), 158–163.

Voldman, S.H. (2004). *ESD: Physics and Devices*. John Wiley & Sons, Ltd, Chichester.

Weeks, J.D., Tully, J.C., and Kimerling, L.C. (1975). Theory of recombination-enhanced defect reactions in semiconductors. *Phys. Rev. B*, 12(8), 3286–3292.

Yamaguchi, M., Ando, K., Yamamoto, A., and Uemura, C. (1984a). Minority-carrier injection annealing of electron irradiation-induced defects in InP solar cells. *Appl. Phys. Lett.*, 44(4), 432–434.

Yamaguchi, M., Ando, K., and Uemura, C. (1984b). Carrier concentration effects on radiation damage in InP. *J. Appl. Phys.*, 55(8), 3160–3162.

Yamaguchi, M., Umemura, C., and Yamamoto, A. (1984c). Radiation damage in InP single crystals and solar cells. *J. Appl. Phys.*, 55(6), 1429–1436.

Ziegler, M., Talalaev, V., Tomm, J.W., Elsaesser, T., Ressel, P., Sumpf, B., and Erbert, G. (2008a). Surface recombination and facet heating in high-power diode lasers. *Appl. Phys. Lett.*, 92(20), 203506.

Ziegler, M., Tomm, J.W., Elsaesser, T., Matthiesen, C., Bou Sanayeh, M. and Brick, P. (2008b). Real-time thermal imaging of catastrophic optical damage in red-emitting high-power diode lasers. *Appl. Phys. Lett.*, 92, 103514.

2

Multi-Component Model for Semiconductor Laser Degradation

The physical mechanisms underlying semiconductor laser degradation are complicated and nonlinear. The observed aging behavior often possesses multiple failure modes that exhibit composite nonlinear trends over aging time. This complicated aging behavior leads to many proposed device-specific and failure mode–specific aging laws that are, in large measure, empirical in nature and often contradictory. In this chapter, we review a multicomponent degradation model that incorporates a Verhulst–Pearl view of defect generation into the physical law, that describes the threshold level of lasers. The outcome of this innovative approach establishes a unified model that explains the observed linear and nonlinear degradation modes, and provides a more organized understanding of the physics behind device degradation. Moreover, it helps unravel composite nonlinear degradation behavior and identify common driving mechanisms in many seemingly unrelated degradation behaviors.

2.1. Introduction

Semiconductor lasers are indispensable and ubiquitous in modern optical communication systems. Estimating the reliability of semiconductor lasers through accelerated life testing (ALT) is of vital importance to optical communication system performance (Joyce *et al.* 1985; Fukuda 1991; Ueda 1996, 2010). The primary objective of this chapter is to review a multicomponent saturable laser degradation model (MCM) which was one of the first attempts to bridge physical degradation mechanisms with experimental observations, regarding rapid and gradual degradations (Lam *et al.* 2003a, 2004).

Chapter written by Samuel K.K. LAM and Daniel T. CASSIDY.

In the study of semiconductor laser degradation, threshold current is a fundamental degradation indicator for semiconductor lasers during ALT (Hakki *et al.* 1985; Sim 1990; Fukuda 1991; Nash 1993; Ohring 1998; Ueda 2010). Modeling of the threshold current with aging time has always been challenging owing to the nonlinear nature and interplay of the degradation mechanisms. Traditionally, the modeling of the threshold current change with aging time has been accomplished by empirical formulae or semiempirical formulae (Hakki *et al.* 1985; Sim 1990; Telcordia Technologies 1998). The impetus for creating a physical degradation model came from a desire to explain threshold degradation observations under one versatile principle. Our physical degradation model comprises two important concepts: a saturable defect growth model and a multicomponent degradation process. The saturable defect growth model, based on the Verhulst–Pearl biological population model (Verhulst 1838; Pearl and Reed 1920), traces a logistic curve that sets out from an initial defect population, which is wafer and process dependent (grown in during epitaxy), and develops into a final state with a maximum finite defect population, which signifies a limited environment that may be governed by the design of the device. The resulting defect model has sufficient versatility – demonstrated later in the chapter – to explain many different observed degradation trends, such as linear, exponential and logistic. This synergized theory between solid-state physics and mathematical biology has the ability to enhance understanding in device reliability.

Through our own research, we found it necessary to introduce the concept of multiple concurrent failure modes to explain more complicated behavior across the spectrum of degradation phenomena. Analogous to the spirit of Fourier analysis, it is our vision that any degradation curves can be broken down into multiple simple logistic processes, characterized by different activation energies. The different activation energies become synonymous to different failure modes of degradation (Nash *et al.* 1985). Comparison between the MCM fit and other existing degradation models (Lam *et al.* 2003a, 2004) shows that the improvement brought about by the MCM is profound, proving that multiple failure modes are at work in the degradation process.

2.2. The physical explanation for saturable degradation

MCM is applicable to both rapid and gradual degradations. Rapid degradation is identified by a rapid increase in threshold current or operating current in the first hundreds of hours of ALT, whereas gradual linear degradation is usually observed after rapid degradation (Horikoshi *et al.* 1979; Hakki *et al.* 1985; Fukuda 1991; Ueda 1996, 2010). Nonetheless, reliable lasers that are free of defects will exhibit only gradual linear degradation, without signs of rapid degradation (Fukuda *et al.* 1983). We show in section 2.4 that under MCM, the difference between a rapid degradation and gradual degradation lies in the activation energy of the intrinsic rate.

Rapid degradation is indicative of the formation of dark-line defects (DLDs) and dark-spot defects (DSDs) (Ueda et al. 1977; Fukuda 1991; Ueda 1996; Jimenez 2003; Hausler et al. 2008; Ueda 2010). DLDs are dislocation networks that form with the help of recombination-enhanced defect climb (REDC) and recombination-enhanced defect glide (REDG) (Ueda 1996), and are likely decorated with condensed impurities. DSDs are precipitation defects by host atoms and sometimes by metal from electrodes. There exist two models for REDC: the extrinsic defect model and the intrinsic defect model. In the extrinsic defect model, dislocations develop by absorbing interstitials (Petroff and Kimerling 1976), whereas in the intrinsic defect model, dislocations form by emitting vacancies (O'Hara et al. 1977). REDG is stress induced (Kamejima et al. 1977). Utilizing the degree of polarization (DOP) technology (Colbourne and Cassidy 1993; Cassidy 2002; Cassidy et al. 2004, 2014), Lisak et al. (2001) show a remarkable correlation between bonding stress and performance degradation in high power diode lasers.

One interesting common observation regarding rapid degradation is that it saturates (Kobayashi and Furukawa 1979; Horikoshi et al. 1979; Fukuda et al. 1983; Hakki et al. 1985; Fatt 1991). This observation lends support to the concept of maximum attainable defect population as a fundamental intrinsic constant in MCM (see M_n in section 2.3). Fukuda et al. (1983) report that the populations of DLDs and DSDs reach a saturation point as the threshold degradation saturates. They remark that not all lasers develop DSD/DLD defects; those lasers that are free of DSD/DLD defects manifest a constant threshold current for over 3,000 h of aging. Coinciding with MCM, the observations suggest that finite numbers of defects are created in each device during wafer growth and processing. By virtue of the random nature of defect creation during growth and processing, one would expect some lasers will have more defects and experience rapid degradation, whereas some lasers will have very few defects and may only experience gradual degradation. These defects are charged and mobile under forward current injection. The process of aging drives these charged mobile defects toward the p-n junction (Kobayashi and Furukawa 1979; Fatt 1991). As the mobile defects accumulate at the p-n junction, they form DLDs (via REDC and REDG) and DSDs via recombination-enhanced defect motion; the threshold current increases rapidly as the DLDs and DSDs continue to multiply. However, the DLD/DSD growth does not continue unabated, because the supply of these mobile defects is limited. We envision that the saturation of laser degradation is realized once all the mobile defects have migrated to the p-n junction. It is also conceivable that limited-in-number mobile defects move toward the active region under the motive force of a non-uniform strain field and condense near the active region and thus give rise to a saturable degradation.

Gradual degradation is believed to be caused by native point defects and/or defect complexes (Ueda 1996, 2010). Device life is typically estimated by extrapolating the gradual degradation (Hakki et al. 1985; Ueda 1996). Ueda proposes that the point defect complexes that are responsible for gradual degradation are created by a recombination-enhanced defect reaction (REDR) process. REDR is a mechanism whereby non-radiative recombination of the injected carriers releases vibrational energy that can create new point defects and/or move existing defects (Kimerling 1978). In gradual degradation, deep-level defects act as non-radiative recombination centers that create vacancy-interstitial point defects assisted by REDR. These point defects can be used to form larger defect complexes. With an assumption that such defect formation cannot continue indefinitely, we believe that gradual degradation will eventually saturate in a logistic process. Moreover, we propose in section 2.5 that the gradual degradation curve that appears to increase linearly is an initial start of a logistic curve, characterized by a large activation energy; thus it takes a long time span to reveal saturation behavior.

2.3. Rate equation for saturable defect population

Laser degradation is controlled by the defect population in its active medium (Petroff and Hartman 1973; Kobayashi and Furukawa 1979; Horikoshi et al. 1979; Dow and Allen 1982; Fukuda et al. 1983; Chuang et al. 1997). Measurements of threshold current over the course of an ALT serves as an accurate indicator for degradation performance. An increase in threshold current signifies an increase in non-radiative recombination at defect sites or at the surface (Fukuda 1991; Coldren et al. 2012; Lam et al. 2005), and hence an increase in the rate of carrier capture at defects in the active region. This increase might be the result of the formation of more non-radiative centers (NRCs), such as precipitation of interstitials and impurities at the core of dislocations, or simply the increase in surface area of existing precipitates. It is interesting to note that the interstitials and impurities are pre-existing defects in the semiconductor die, i.e. defects that were created at fabrication of the semiconductor die. Thus, one would expect the behavior of saturable degradation to be wafer dependent. We proposed that defect growth, analogous to other biological population growth, is logistic in nature and is saturable given ample time for its development (Lam et al. 2003a).

To derive the defect population as a function of time t, we begin by describing the defect population $N(t)$ at time t using a rate equation that has a first term that represents a growth rate proportional to the defect population $N(t)$ and a second growth limiting term that is quadratic in the defect population $N(t)$ (Maki and Thompson 1973; Dym and Ivey 1980):

$$\frac{\mathrm{d}}{\mathrm{d}t}N(t) = K_n N(t) - C_n N(t)[N(t) - 1] \qquad [2.1]$$

Without the second term, K_n is simply the rate in the case of an unlimited supply of materials for growth. The second term causes the growth to saturate as the population increases; its origin arises from the observation that the saturation rate is proportional to the number of paired interactions between N defects, which is $N(N-1)/2$. The rates K_n and C_n represent a degradation process driven by the electron-hole recombination at the defect sites (Chuang *et al.* 1997). Inasmuch as degradation always involves diffusion at the atomic level, it is customary to incorporate the Arrhenius relation in the thermally activated rates utilized in our model. For example, a rate, $X(t)$, set in the Arrhenius form, becomes (Hartman and Dixon 1975; Joyce *et al.* 1985; Nelson 1990; Klinger *et al.* 1990; Fukuda 1991; Nash 1993; Tobias and Trindade 1995; Lewis 1996; Ohring 1998; Chang *et al.* 2012)

$$X(T) = X_\infty e^{-E_a/(kT)} \qquad [2.2]$$

X_∞ denotes $X(T = \infty)$, E_a is the activation energy in eV, k is the Boltzmann constant in eV/K and T is the absolute temperature in K. Combining electron-hole recombination-assisted rate processes and Arrhenius relations yields the rates K_n and C_n as follows:

$$K_n = \kappa n p \, e^{-E_a/(kT)} = \kappa n^2 e^{-E_a/(kT)} \qquad [2.3]$$

and

$$C_n = c n p \, e^{-E_a/(kT)} = c n^2 e^{-E_a/(kT)} \qquad [2.4]$$

where n is the electron concentration and p is the hole concentration. The $\kappa n p$ and $c n p$ products are recombination-enhanced defect generation terms introduced by Chuang *et al.* (1997). For an undoped active region, $p = n$ is assumed. The subscript n in K_n and C_n signifies that these rates are driven by electron-hole recombination that, in turn, depends on the driving current, a stress parameter in ALT.

For device reliability studies, it is customary to describe the defect population in terms of its defect density by replacing N with $N_d V$, where N_d and V denote the defect density and the volume of the active region, respectively. We shall define a new variable, $M_n = (\kappa + c)/(cV)$, and the benefit of doing this will become clear later as we discuss saturation. Equation [2.1] can then be simplified to

$$\frac{\mathrm{d}}{\mathrm{d}t} N_d(t) = C_n V M_n N_d(t) \left[1 - \frac{N_d(t)}{M_n} \right] \qquad [2.5]$$

with solution

$$N_d(t) = \frac{N_{d0} M_n}{(M_n - N_{d0}) e^{-C_n V M_n t} + N_{d0}} \qquad [2.6]$$

where $N_{d0} = N_d(0)$ is the initial population and $C_n V M_n$ is the electron-hole recombination-driven intrinsic growth rate (also known as per capita growth rate).

We shall refer to $C_n V M_n$ as the intrinsic growth rate to differentiate it from the population growth rate, $dN_d(t)/dt$. Equation [2.6] is a logistic function known as the Verhulst–Pearl equation (Verhulst 1838; Pearl and Reed 1920) that has its origin in mathematical biology. The formatting of equation [2.5] demonstrates how the term $1 - N_d(t)/M_n$ helps achieve saturation, i.e. as $N_d(t)$ reaches M_n, $dN_d(t)/dt$ depletes to zero and hence growth ceases at this point. Biologists see $1 - N_d(t)/M_n$ as a measure of vacant places available for growth (Gause 1971). Including $1 - N_d(t)/M_n$ as a multiplicative factor in the rate term is equivalent to placing the population in a finite growth environment. It is self-evident that M_n is none other than a maximum attainable defect density, the physical quantity that defines saturation.

The physical insight provided by equation [2.6] is significant. At the beginning of aging, the defect population is much less than M_n. The curve grows, analogous to an exponential curve, with a rate of $C_n V M_n$. The rate of growth of defect population reaches a global maximum at the inflection point of equation [2.6], which can be shown to be $M_n/2$, half of the maximum attainable defect population of the system. These physical characteristics permit researchers to extract defect density information through numerical fitting to experimental data: a capability that is unique to our physical model and is not otherwise available in other existing models, to our best knowledge.

In order to provide more insight on how equation [2.6] works, we plot the equation in Figure 2.1a assuming the following starting values for its parameters: $C_n V M_n = 1$, $M_n = 100$, and $N_d(0) = 20$. The black curve of Figure 2.1 shows the unlimited case when the second factor $1 - N_d(t)/M_n = 1$, then $N_d(t) = N_{d0} \exp(C_n V M_n t)$, which is an exponential growth in time. The parameter M_n gives the sustainable population. In the limit as $t \to \infty$, $N_d(t \to \infty) = M_n$. For the blue curve, the sustainable population $N_d(\infty) = M_n = 100$, whereas the green curve, by contrast, has only half of the blue curve's sustainable population. Note, for small t, that $N_d(t)$ increases exponentially with time t and that the saturation level is proportional to M_n. The red curve is for $C_n = 0$. For the choice of parameters selected to plot the red curve, the rate equation is $dN_d(t)/dt = 0$ and $N_d(t) = N_d$ is a constant for all time. The brown curve plots the solution for $M_n = 1 < N_{d0}$. Clearly different behavior is obtained for different values of intrinsic growth rate, $C_n V M_n$.

Also of interest is the population growth rate dN_d/dt as plotted in Figure 2.1b. The curve illustrates one unique property of a Verhulst–Pearl growth curve. At the beginning, the population growth rate increases steadily when resources for growth are abundant. Then it reaches a peak when the saturation factor $1 - N_d(t)/M_n$ starts to overcome the growth trend. By varying the intrinsic growth rate $C_n V M_n$, one will immediately realize that the maximum population growth rate dN_d/dt always occurs at $M_n/2$ regardless of the values of the initial population $N_d(0)$ and the intrinsic growth rate. Once the population surpasses half of the maximum sustainable population M_n, the population growth rate drops steadily and eventually stops when

$N_d(t) = M_n$. We can derive the time to reach half of the maximum sustainable population by substituting $N_d(t) = M_n/2$ in equation [2.6]. In so doing, we obtain the following expression:

$$t_{1/2} = \frac{1}{C_n V M_n} \ln\left(\frac{M_n}{N_{d0}} - 1\right) \qquad [2.7]$$

where $t_{1/2}$ denotes the time when the population reaches half of its maximum capacity.

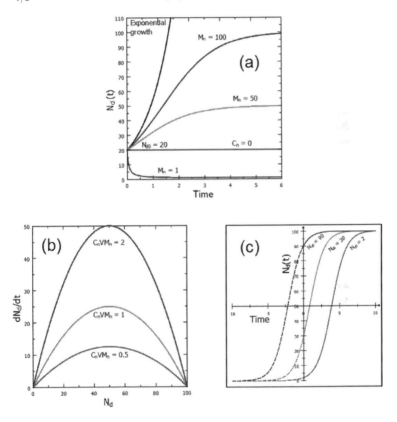

Figure 2.1. (a) Plots of equation [2.6] with different values of M_n and C_n; (b) growth rate with different values of $C_n V M_n$; (c) plots of equation [2.6] with $N_{d0} = 90$, 30 and 2. For a color version of this figure, see www.iste.co.uk/vanzi/reliability.zip

Another very useful property of the Verhulst–Pearl equation is illustrated in Figure 2.1c. It demonstrates that equation [2.6] is antisymmetric about $N_d(t_{1/2}) = M_n/2$ regardless of the value of N_{d0}. Consider three plots of equation [2.6] with $C_n V M_n = 1$ and $M_n = 100$ as shown in Figure 2.1c. The blue curve has

$N_{d0} = 90$, the green curve has $N_{d0} = 30$ and the red curve has $N_{d0} = 2$. We have extended the curves into the negative value of time (dashed curve as opposed to solid curve) only to show the complete symmetry of the curves about $N_d(t) = M_n/2$. Note that these curves are basically identical but shifted horizontally so that each curve will have a corresponding N_{d0} correctly set at $t = 0$. In other words, the value of N_{d0} relative to M_n defines the initial curvature (or rate of growth) of the aging curve. Aging curves reported in the literature can exhibit either an initial exponential increase in the early stage Hakki *et al.* (1985) call it the incubation period as does the red curve or an initial linear increase (Horikoshi *et al.* 1979) as does the green curve. Using equation [2.6], the difference between the two observations can be justified by the difference in N_{d0}. For example, if N_{d0} is close to $M_n/2$, an initial linear increase will be expected, whereas if N_{d0} is much less than $M_n/2$, an initial exponential increase will be observed. Finally, for $N_{d0} > M_n/2$, the growth curve (the blue curve) for $t > 0$ has a shape resembling

$$N_d(t) \propto t^m \qquad [2.8]$$

where t is the aging time and m is an empirical constant. Our simple demonstration in the context of N_{d0} explains the origin for the t^m dependence of the popular empirical equation for the threshold current (Sim 1990; Fukuda 1991; Okayasu and Fukuda 1992; Telcordia Technologies 1998),

$$\Delta I_{th}(t)/I_0 = A t^m \exp(-E_a/(kT)) \qquad [2.9]$$

where:

$I_0 = I_{th}(0)$;

A and m are empirical constants;

E_a is the activation energy;

k is the Boltzmann constant;

T is the absolute temperature;

t is the aging time.

2.4. Saturable laser degradation by single defect population

The concept of saturable laser degradation has been known for over 30 years (Horikoshi *et al.* 1979; Hakki *et al.* 1985). Assuming that the saturable degradation is attributed to an increase in non-radiative current, the change over time t in threshold current can be written as

$$\Delta I_{th}(t) = I_{nr}(t) - I_{nr}(0), \qquad [2.10]$$

where $I_{nr}(t)$ is the non-radiative current and

$$I_{nr}(t) = \frac{q V A N_d(t) n_{th}}{\eta_i}, \qquad [2.11]$$

q is a single electron charge, V is the volume of active region and η_i is the internal injection efficiency. $A N_d(t) n_{th}$ is the non-radiative carrier capture rate of the defects at the initial threshold carrier density n_{th} (Chuang et al. 1997).

With $N_d(t)$ as derived in equation [2.6], we arrive at

$$\Delta I_{th}(t) = \frac{q V A n_{th}}{\eta_i} \left[\frac{M_n N_{d0}}{N_{d0} + (M_n - N_{d0}) \exp(-C_n V M_n t)} - N_{d0} \right] \qquad [2.12]$$

where C_n is the temperature-dependent, electron-hole recombination-driven rate shown in equation [2.4]. Equation [2.12] describes the threshold change for a single defect population. A typical saturable degradation curve, as plotted by equation [2.12], is shown in Figure 2.2. The curve starts at $q V A n_{th} N_{d0}/\eta_i$ and saturates at $q V A n_{th} M_n/\eta_i$. The inflection point of the curve can be shown to be at $N_d(t) = M_n/2$. The segment of the aging curve from the beginning of aging time to the inflection point marks the incubation period (Hakki et al. 1985).

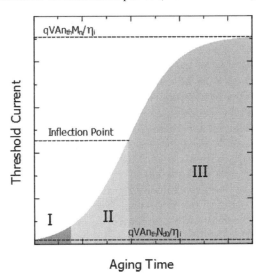

Figure 2.2. *A single component threshold current degradation curve. Region I: purple; region II: yellow; region III: green. For a color version of this figure, see www.iste.co.uk/vanzi/reliability.zip*

Degradation reported in literature does not always exhibit saturation. We believe such a discrepancy is attributed to the practice of censoring during ALT. Using time-censored and failure-censored ALT as examples for our argument, ALT is terminated either at a predetermined time or when a predetermined number of failures has been reached (other censoring schemes exist but are not important in our present discussion) (Tobias and Trindade 1995). In time-censored ALT, many diode lasers often survive the entire test. Many of these diode lasers tend to exhibit wear-out degradation that takes too long to reveal the saturable behavior over practical time spans. In failure-censored ALT, failure is often concluded when the threshold current reaches an unsustainable level, which may happen long before the onset of saturation. This type of failure criteria leaves very few benefits and incentives for diode laser companies to extend their device study long past their typical duration to elucidate the complete wear-out degradation behavior.

How censoring affects our interpretation of the extrapolated behavior of the degradation can be demonstrated by dividing the saturable curve in Figure 2.2 into three sections (I, II and III). In many practical ALT, failures do not occur during the aging period and extrapolation will be utilized to assess failure time. Depending on the censoring time, a test operator can arrive at different conclusions of a projected lifetime of a device. For instance, if an operator decides to stop the experiment right at the end of region I of the curve in Figure 2.2, the curve may be interpreted as linear. Extrapolation at this point can underestimate the degradation if the projected time is not far ahead or overestimate the degradation if the projected time is set beyond the saturation point. Moreover, it also could complicate determination of activation energy (Lam *et al.* 2003b). If the censoring occurs at the end of region II, the inflection point of the curve in Figure 2.2, it can be interpreted as a rapid degradation (Ueda 2010). Extrapolation from this point will deviate profoundly from the true value. A more sensible approach is to perform extrapolation with data well in region III. Extrapolation performed from points in region III, especially after the sign of saturation, will lead to more accurate predictions.

Activation energy associated with the defect growth rate also can mislead test operators, similar to the aforementioned example. For degradation curves that manifest multiple components, each component or failure mode degrades at a different rate in accordance with its activation energy. According to the Arrhenius relation in equation [2.2], failure modes associated with smaller activation energies drive the rapid degradation. Because thermally activated rates are proportional to $\exp(-E_a/(kT))$, the degradation rate becomes larger for smaller E_a. By the same token, larger E_a is associated with gradual degradation, because the rate is smaller for larger E_a. Figure 2.3 demonstrates the Arrhenius effect in a simulation of three single-component threshold degradation curves with different activation energies extracted from the same test period. The width of the abscissa is the length of the censored ALT. These three simulated curves only differed by the activation energy. For smaller activation, such as the blue curve with E_{a1}, the entire growth curve is clearly visible to the operator. This is the ideal case whereby all degradation information is present. When the activation energy is larger, such as the green curve, the censored test window starts to clip the degradation curve. Some information of the degradation is obscured, but extrapolation can probably provide a reasonable estimate to the true value. Finally, for an activation energy that is too large for the censored test window to cover its full development (as shown in the brown curve), the degradation curve appears to be a linear degradation. Such misleading data are possible because the censoring only renders the beginning of an extensive logistic aging curve visible to the test operator. In this case, not enough information is available for extrapolation with a high confidence level.

2.5. Multicomponent model for degradation dynamics

Independent laser degradation studies reported observation of multicomponent degradation curves (Lam *et al.* 2003a, 2004; Horikoshi *et al.* 1979; Hakki *et al.* 1985; Kobayashi and Furukawa 1979; Nash *et al.* 1985; Fatt 1991). The origin of multiple components seems to be a universal phenomenon across different types of diode lasers. Our experimental data, which resemble a superposition of multiple saturable degradation curves, prompted us to propose an MCM for laser degradation, in conjunction with the defect model derived in equation [2.12] (Lam *et al.* 2003a). In an MCM, non-radiative defect complexes are led by different species that are classified by their activation energies, E_a. As shown in Figure 2.3, defects that have low activation energies tend to saturate faster and are associated with fast degradation. In contrast, defects that have high activation energies saturate slower and are associated with slow degradation.

In deriving the MCM, we assume a non-interacting growth of multiple defect complexes. This assumption leads to ζ different rate equations for describing the growth of ζ different defects N_{dm} (Lam *et al.* 2003a):

$$\frac{dN_{d1}(t)}{dt} = N_{d1}(t)\left[a_{11} - a_{12}V N_{d1}\right]$$

$$\frac{dN_{d2}(t)}{dt} = N_{d2}(t)\left[a_{21} - a_{23}V N_{d2}\right]$$

$$\vdots \qquad\qquad\qquad\qquad\qquad [2.13]$$

$$\frac{dN_{d\zeta}(t)}{dt} = N_{d\zeta}(t)\left[a_{\zeta 1} - a_{\zeta,\zeta+1}V N_{d\zeta}\right]$$

where

$$a_{m,1} = (\kappa_m + c)\, n^2 \exp[-E_{am}/(kT)]$$

$$a_{m,m+1} = c\, n^2 e^{-E_{am}/(kT)}$$

$$m = 1,2,...\zeta$$

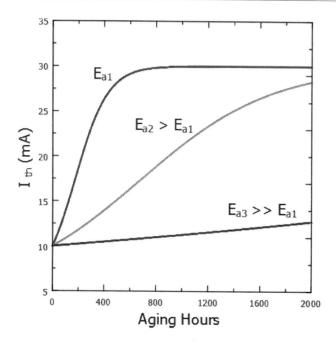

Figure 2.3. *Simulation of degradation curves with respect to different activation energy E_a. Blue: E_{a1}; green: $E_{a2} > E_{a1}$; brown: $E_{a3} >> E_{a1}$. For a color version of this figure, see www.iste.co.uk/vanzi/reliability.zip*

Note that $c_m = c$ is assumed. This assumption imposes a rule that any two $a_{m,m+1}$ will be identical if their activation energies E_{am} are equal. Each solution in equation [2.13] has the same form as equation [2.6] with

$$N_d(t) = N_{d1}(t) + N_{d2}(t) + \ldots + N_{d\zeta}(t).$$ [2.14]

The threshold degradation equation in MCM becomes

$$\Delta I_{th}(t) = \frac{qVAn_{th}}{\eta_i}$$

$$\times \left[\sum_{m=1}^{\zeta} \frac{M_{nm} N_{dm}(0)}{N_{dm}(0) + (M_{nm} - N_{dm}(0)) \exp(-C_{nm} V M_{nm} t)} - N_{dm}(0) \right]$$ [2.15]

where $M_{nm} = a_{m,1}/a_{m,m+1}$ is the maximum defect density for the mth component. The intrinsic rates C_{nm} for each component are defined as

$$C_{nm} = a_{m,m+1} = cn^2 e^{-E_{am}/(kT)}$$ [2.16]

Figure 2.4 shows the aging data and its MCM fitted curve. In essence, equation [2.15] describes the following defect growth scenarios. Defect complexes form and grow by consuming mobile defects that are driven to the active region. The fastest growing defect complexes deplete all the mobile defects in their vicinities, then ceases further development and forms the first "knee" in the aging curve. The slower growing defect complexes continue to grow until their sources of mobile defects are exhausted. These scenarios explain how multiple "knees" of saturation can be observed.

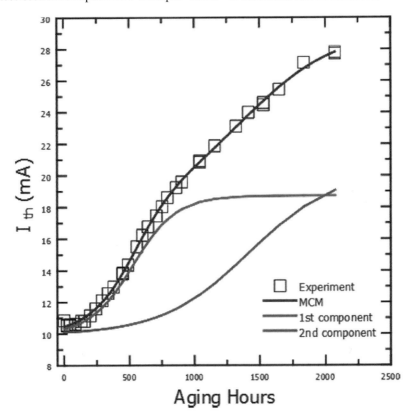

Figure 2.4. *A two-component MCM model fitted to experimental degradation data. Black: best fit two-component curve; purple: first component; blue: second component. For a color version of this figure, see www.iste.co.uk/vanzi/reliability.zip*

2.6. Annealing effect

Annealing is the result of defect annihilation (Kobayashi and Furukawa 1979; Horikoshi *et al.* 1979; Fatt 1991; Yamada *et al.* 1991; Fukuda 1991; Zhang *et al.* 1992) and reduction of internal loss (Werner *et al.* 1988; Jalonen *et al.* 1997) upon aging. What happens microscopically that causes defect annealing is not entirely understood. To model the annealing effect with normal degradation, we propose the following expression for the change in threshold:

$$\Delta I_{th}(t) = I_{nr}(t) - I_{nr}(0) + B\left[n_{th}^2(t) - n_{th}^2(0)\right]$$ [2.17]

Equation [2.17] is a second-order power series approximation for the change in threshold in n_{th} where B is a power series expansion coefficient and $I_{nr}(t)$ is given by equation [2.15]. The term $B[n_{th}^2(t) - n_{th}^2(0)]$ represents the collective annealing effect in Auger, surface, leakage and amplified spontaneous emission. In MCM, annealing is manifested by the change of internal loss α_i, because any change of internal loss can be detected in both the threshold current and the output efficiency. Other effects such as defect annihilation probably exist, but the effects are absorbed into the saturable degradation and cannot be easily separated from the concurrent degradation process. Generally, the threshold carrier density is related to the internal loss (Coldren *et al.* 2012) in a logarithmic form that is assumed for quantum well devices (DeTemple and Herzinger 1993):

$$\Gamma g_0 \ln\left(\frac{n_{th}}{n_{tr}}\right) = \alpha_i + \alpha_m$$ [2.18]

where Γ is the optical confinement factor, g_0 is a gain coefficient, n_{tr} is the carrier density at transparency and α_m is the mirror loss. The internal loss α_i is associated with optical defects (ODs) that scatter and absorb light quanta in the active region, as opposed to non-radiative recombination defect centers. The internal loss affects the output intensity, slope efficiency and threshold carrier density, but non-radiative recombination defect centers only directly affect the threshold current. In essence, all physical mechanisms that change the optical loss are combined in one variable, α_i. Henceforth, the reduction of internal loss as a result of annealing will be referred to as OD annealing.

The time evolution of internal loss is the sum of a degradation term $\alpha_1(t)$ and an annealing term $\alpha_2(t)$

$$\alpha_i = \alpha_0 + \alpha_1(t) + \alpha_2(t).$$ [2.19]

The first term α_0 is the residual internal loss at $t = 0$ that does not change with aging. The second term $\alpha_1(t)$ represents the degradation term but is assumed to be negligible. The third term $\alpha_2(t)$ represents the annealing term. We postulate that the annealing is a reverse process of saturable defect generation with the form similar to equation [2.5] (Lam *et al.* 2004). The proposed rate equation for the annealing term is

$$\frac{d\alpha_2(t)}{dt} = -RL\alpha_{i,max}\,\alpha_2(t)\left(1 - \frac{\alpha_2(t)}{\alpha_{i,max}}\right).$$ [2.20]

The parameter $\alpha_{i,max}$ denotes the maximum reducible internal loss in the annealing process. The rate at which the annealing will saturate is governed by R and $L\alpha_{i,max}$. The rate equation has a solution (Lam *et al.* 2004)

$$\alpha_i(t) = \alpha_0 + \alpha_{i,max} - \frac{\alpha_{i,max}\, \alpha_{init}}{\alpha_{init} + (\alpha_{i,max} - \alpha_{init})\, \exp(-RL\alpha_{i,max}\, t)} \qquad [2.21]$$

where

$$R = R_0\, e^{-E_{a\alpha}/(kT)}. \qquad [2.22]$$

R_0 is the rate of reduction of the internal loss as temperature goes to infinity. The thermal activation energy for R is named $E_{a\alpha}$ to distinguish it from other activation energies used in equation [2.13]. The threshold current density can then be derived as follows:

$$n_{th}(t) = n_{tr}\, \exp\left(\frac{\alpha_0 + \alpha_m}{\Gamma g_0}\right)$$

$$\times \exp\left[\frac{1}{\Gamma g_0}\left(\alpha_{i,max} - \frac{\alpha_{i,max}\, \alpha_{init}}{\alpha_{init} + (\alpha_{i,max} - \alpha_{init})\, \exp(-RL\alpha_{i,max}\, t)}\right)\right]. \qquad [2.23]$$

The change in threshold current with annealing can be written as:

$$\Delta I_{th}(t) = B\left[n_{th}^2(t) - n_{th}^2(0)\right]$$

$$+ \frac{qVA}{\eta_i}\left[\sum_{m=1}^{\zeta} \frac{M_{nm}N_{dm}(0)\, n_{th}(t)}{N_{dm}(0) + (M_{nm} - N_{dm}(0))\, \exp(-C_{nm}VM_{nm}t)} - N_{dm}(0)n_{th}(0)\right]. \qquad [2.24]$$

Figure 2.5a shows an example of a threshold aging curve fitted by the extended MCM. The corresponding OD annealing and degradation processes are plotted in Figures 2.5b and 2.5c, respectively. The OD annealing component encompasses the collective effect of the annealing-induced change by Auger, surface, leakage and amplified spontaneous emission, all of which depend on α_i through n_{th}.

Diode laser physics requires that the differential quantum efficiency η_d depends on the internal loss by the following relation (Agrawal and Dutta 1993; Coldren *et al.* 2012):

$$\eta_d = \eta_i \frac{\alpha_m}{\alpha_i + \alpha_m}. \qquad [2.25]$$

Consequently, any change in α_i (or α_m) should be readily observable in differential quantum efficiency measurements. Because the internal quantum efficiency η_i represents the fraction of the carriers in the applied current that ends up in the active region, we assume it does not change during aging (Kallstenius *et al.* 2000). In essence, annealing observed in both the threshold current and the output efficiency should be driven by the change in α_i. Figure 2.6a shows the experimental output efficiency data of the same device in Figure 2.5, exhibiting OD annealing in the first 1000–1500 h as expected from equation [2.25]. Figures 2.6b and 2.6c show the annealing and degradation components, respectively.

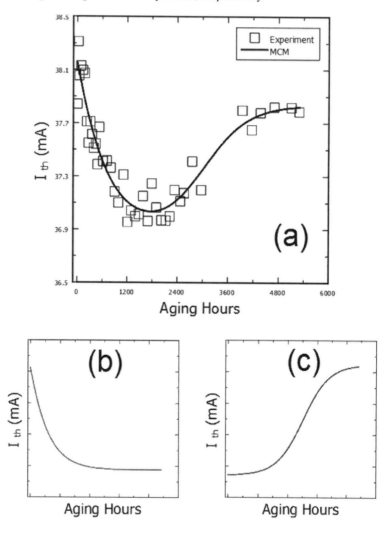

Figure 2.5. *(a) MCM model fitted to a partially annealed threshold aging curve; (b) red: the annealing component; (c) blue: the degradation component. For a color version of this figure, see www.iste.co.uk/vanzi/reliability.zip*

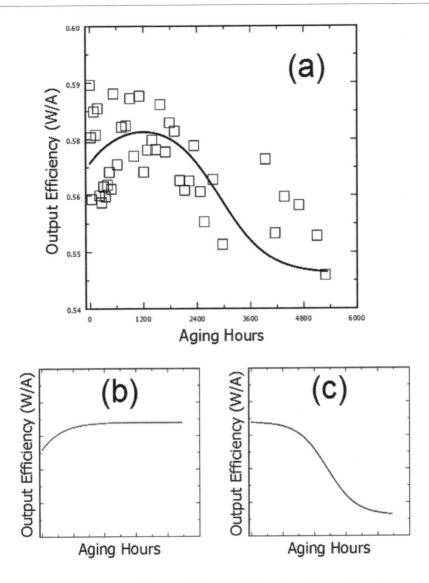

Figure 2.6. (a) Output efficiency exhibiting sign of annealing;
(b) red: the annealing component; (c) blue: the degradation component.
For a color version of this figure, see www.iste.co.uk/vanzi/reliability.zip

Thus far, our discussion on annealing has only focused on OD annealing, i.e. the reduction of internal loss. The mathematical framework of the MCM can support both OD annealing and defect annihilation. We conclude this section by deriving two hypothetical MCM models with defect annihilation that are of interest. In the first hypothetical model, we assume that defect annihilation takes the form of $\alpha I^\beta N_d(t)$ in accordance with various studies (Horikoshi et al. 1979; Kobayashi and Furukawa 1979; Fatt 1991). The variable α is a proportionality constant, I denotes the stress current and β is the factor that enables a nonlinear dependence on the stress current. Defect growth and defect annihilation occur concurrently in the same population. The defect rate equation in equation [2.5] can then be rewritten as

$$\frac{d}{dt} N_d(t) = N_d(t) \left[K_n + C_n - \alpha I^\beta \right] - C_n V N_d^2(t).$$ [2.26]

where $-\alpha I^\beta$ is the defect annihilation rate and is assumed to obey the Arrhenius relation, as shown in equation [2.2]. The resulting threshold aging curve retains the form of equation [2.12]

$$\Delta I_{th}(t) = \frac{q V A n_{th}}{\eta_i} \left[\frac{M_n N_{d0}}{N_{d0} + (M_n - N_{d0}) \exp(-C_n V M_n t)} - N_{d0} \right]$$ [2.27]

where

$$M_n = \frac{K_n + C_n - \alpha I^\beta}{C_n V}$$

$$K_n = \kappa n^2 e^{-E_a/(kT)}$$

$$C_n = c n^2 e^{-E_a/(kT)}$$

The subtle difference in the threshold aging curve is embedded in M_n. The maximum sustainable defect population term M_n now has a new offset attributable to the defect annihilation process, $\alpha I^\beta/(C_n V)$. Generally, when annealing competes with degradation in the same defect population, it is hard to observe annealing unless $\alpha I^\beta > (K_n + C_n)$ is true (Lam et al. 2004).

In contrast to the first hypothetical model, the second hypothetical MCM model with defect annihilation assumes the defect degeneration and annihilation occur independently. This is a case that mirrors the annealing phenomena reported by Fatt (1991), Kobayashi and Furukawa (1979) and Horikoshi et al. (1979). In this case, annihilation is the reverse process of a logistic growth. The resulting model can be shown to be (Lam et al. 2004)

$$\Delta I_{th}(t) = \frac{q V A n_{th}}{\eta_i} \left[\frac{M_n N_d(0)}{N_d(0) + (M_n - N_d(0)) \exp(-C_n V M_n t)} - N_d(0) \right]$$
$$+ \frac{q V A n_{th}}{\eta_i} \left[N_a'(0) - \frac{M_n' N_a'(0)}{N_a'(0) + (M_n' - N_a'(0)) \exp(-C_n' V M_n' t)} \right].$$ [2.28]

The second half of the equation introduces some new variables pertaining to the annealing process. The maximum attainable annealing is limited by M_n'. $N_a'(0)$ is the initial annealing population and C'_n is the intrinsic annealing-related rate that obeys the Arrhenius relation. As an example, equation [2.28] assumes only one degradation component and one annealing component, as depicted in Figure 2.5. Depending on the situation, more components can be added, as in equation [2.15], if required.

2.7. Guide to MCM applications

Modern diode lasers have typical lifetimes exceeding hundreds of thousands of hours. It is impractical to estimate the lifetimes of these devices at normal operation conditions, as part of an engineering practice in a product lifecycle. For this primary reason, estimation of the longevity of diode lasers is typically performed by ALT. In the procedure, operators age the diode lasers at elevated temperatures and higher drive current. During the course of the aging, the lasers are periodically removed from the test to measure performance, such as threshold current. Operators will evaluate the lasers based on their charted performance. Even though under stress conditions few lasers fail or reach the failure level during ALT, it is customary to extrapolate the results to a failure level as a measure of their lifetimes at a stress level. Therefore, extrapolation or modeling of the aging curves forms an integral part of the device lifetime estimation. In this section, we review the patterns of threshold aging curves. Because of the unconventional approach introduced by MCM, we discuss qualitatively on how to model these aging patterns with the MCM.

In the case of a single component, the threshold current aging curves can be categorized in four general cases: linear, logistic growth (degradation), exponential and logistic decay (annealing) functions, as shown in Figure 2.7. Moreover, these four cases can all be explained by MCM: Figure 2.7 is derived by a single-component MCM using different E_a, $N_d(0)$, $C_n V M_n$ and censoring conditions to demonstrate how these four types of aging curves can be modeled by one single equation. Linear degradation (Figure 2.7a) is otherwise known as gradual degradation (Hakki *et al.* 1985; Fukuda 1991; Ueda 2010); for MCM, a linear degradation is the result of a censored logistic curve, equation [2.12], that has a large E_a. Logistic degradation (Figure 2.7b) is considered rapid degradation characterized by a small E_a (Hakki *et al.* 1985; Nash *et al.* 1985). In the past, logistic degradation was observed to exhibit either with an incubation period (Hakki *et al.* 1985) or without (Fukuda *et al.* 1983; Kallstenius *et al.* 2000; Huang 2013). In Figure 2.7b, we show these two subcategories of rapid-degradation-inspired logistic curves in one plot. The difference in these two cases can be explained by equation [2.12] with different $N_d(0)$ (see Figure 2.1 for details). Exponential degradation (Figure 2.7c) is observed in devices that experience rapid degradation (Chuang *et al.* 1997); for MCM, it is a logistic curve, equation [2.12], that is either censored or abruptly terminated by a sudden failure. Logistic annealing curves (Figure 2.7d) have been observed in different diode laser products (Kobayashi and Furukawa 1979; Fatt 1991); it is possible to model a single annealing curve by the annealing term of equations [2.24],

[2.27] or [2.28]. In Figure 2.7d, we identify and trace two subcategories of the logistic annealing curves with different initial annealing populations in one plot.

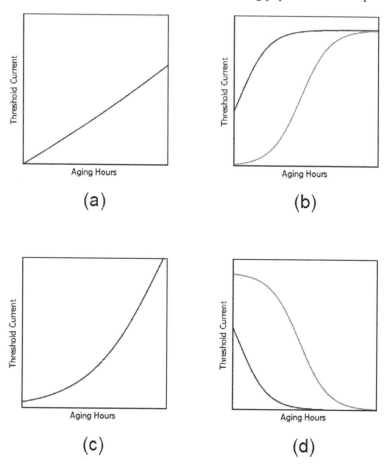

Figure 2.7. *Summary of single-component degradation and annealing curves: (a) linear, (b) logistic growth, (c) exponential and (d) logistic decay. For a color version of this figure, see www.iste.co.uk/vanzi/reliability.zip*

For multicomponent aging patterns, the shapes of the curves are typically a combination of the aforementioned single-component degradation curves. There are four basic combinations: logistic growth-linear, power-linear, multilogistic and logistic decay-logistic growth, as shown in Figure 2.8. Once again, as a demonstration, we recreate these complex degradation patterns – found in the literature – in Figure 2.8 using different E_a, $N_d(0)$, $C_n V M_n$ and censoring conditions. However, in these examples, we only demonstrate two-component cases; more complicated aging curves can certainly be modeled with more components as operators see fit.

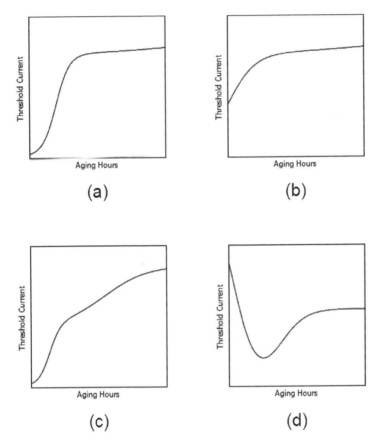

Figure 2.8. *Summary of multicomponent degradation curves: (a) logistic growth-linear, (b) power-linear, (c) multilogistic and (d) logistic decay-logistic growth*

The logistic growth-linear degradation curves in Figure 2.8a are a combination of a transient degradation mode and a gradual linear degradation (Hakki *et al.* 1985; Nash 1993). The transient degradation mode is characterized by an incubation period, and it tends to saturate faster because of its small activation energy. In contrast, the gradual linear degradation mode has a large activation energy. In MCM, we interpret both modes as logistic growth curves – equation [2.15] – with two different activation energies. The power-linear degradation curve in Figure 2.8b is the most popular textbook example (Newman and Ritchie 1973; Horikoshi *et al.* 1979;

Hakki *et al.* 1985; Nash 1993; Ohring 1998; Telcordia Technologies 1998; Huang 2013). In MCM, we treat it the same way as the logistic growth-linear degradation curve – equation [2.15] – except that the logistic growth does not exhibit an incubation period because the size of $N_d(0)$ is of the order of $M_n/2$. Because such a combination makes the curve reminiscent of a power function, equation [2.9] (Sim 1990) becomes the popular heuristic model for this type of aging curve.

Complex aging curves exhibiting multiple "knees" (Lam *et al.* 2003a), as depicted in Figure 2.8c are the result of multiple concurrent logistic degradation modes, equation [2.15], with different E_a. Finally, the last example is the logistic decay-logistic growth degradation curve (Fatt 1991; Lam *et al.* 2004). In MCM, it has a logistic decay component and a logistic growth component. The logistic decay component is an annealing phenomenon as explained in Figure 2.5, whereas the logistic growth is the outcome of a normal saturable degradation as described by equation [2.12]. Equations [2.24], [2.27] or [2.28] will be the choice for modeling this type of aging curve.

The estimated lifetime at the operation condition is typically calculated using the Arrhenius relation (Hartman and Dixon 1975; Eckler 1985; Fukuda 1991; Ohring 1998; Telcordia Technologies 1998):

$$t_f = t_s \exp\left[\frac{E_a}{k}\left(\frac{1}{T_f} - \frac{1}{T_s}\right)\right]$$
[2.29]

where

t_f is the estimated lifetime at the intended operation temperature T_f;

t_s is the measured lifetime at the stress temperature T_s;

E_a is the activation energy;

k is the Boltzmann constant;

T_f is the operation temperature in K;

T_s is the stress temperature in K.

It is well known that the Arrhenius model shown in equation [2.29] is not applicable to multiple chemical reactions that have different activation energies (Escobar and Meeker 2006). Following Nash's approach (Nash *et al.* 1985; Nash 1993), we recommend that ALT should be extended long enough to stabilize all saturable transient modes of degradation and annealing, leaving only the wear-out failure mode for determining the device life. Referring to Figure 2.5, the wear-out failure mode is basically the second or last component of the composite degradation curves, whereas the first component is considered a transient mode. With only the wear-out mode, equation [2.29] will provide a valid lifetime estimation.

As demonstrated in Figure 2.3, the combination of censoring and large activation energy often renders the wear-out mode to appear linear. Attempting to fit such limited data with the logistic curve will only produce inaccurate lifetime estimation. Yet extending the ALT to capture the complete features of the wear-out mode is proved to be impossible in practice. This common dilemma can be resolved by utilizing a linear approximation of equation [2.12]. To demonstrate linear extrapolation of the wear-out failure mode, we consider a hypothetical wear-out aging curve (solid blue) depicted in Figure 2.9. Wear-out modes typically possess a relatively large activation energy, resulting in a small $C_n V M_n t << 1$ for small t. Utilizing the Maclaurin series approximation $e^x = 1 + x$ and $M_n >> N_{d0}$, it can be shown that the early degradation of equation [2.12] can be approximated by a linear relationship as follows:

$$\Delta I_{th}(t) = \frac{qVAn_{th}N_{d0}C_n V M_n t}{\eta_i}. \tag{2.30}$$

This linear curve is depicted by the dashed black line in Figure 2.9.

Figure 2.9. *Hypothetical wear-out failure mode (solid blue) and a linear approximation (dashed black) at the beginning of the failure mode. For a color version of this figure, see www.iste.co.uk/vanzi/reliability.zip*

There are caveats in approximating the wear-out degradation mode with the linear approximation in equation [2.30]. The approximation is valid only when

$C_n V M_n t \ll 1$, which is normally near the beginning of the aging curve. Linear extrapolation tends to be an underestimate, as evident in Figure 2.9. As a result of the underestimation, extrapolating beyond the test data window has to be performed with care.

2.8. Summary

The standard model of aging appears to be that defects create more defects. If the rate of production of defects is proportional to the number of defects, then the rate of growth of defects would be exponential. Exponential growth is not always observed. Saturation of the rate of growth of defects is observed for some non-negligible subset of lasers. Given this observation, we suggest that a model for the degradation of lasers with aging might be growth of non-radiative recombination centers (NRCs), such as DSDs and DLDs, by condensation of existing point defects at NRCs. As the supply of point defects is exhausted by gettering at the NRCs, the rate of growth of the NRCs will saturate. This is analogous to saturation of biological species in the Verhulst–Pearl model of populations that compete for food. The suggested logistic model predicts that changes in threshold, which we take as a proxy for the density of NRCs, will saturate with time for devices that have a limited supply of resources for the growth of defects.

The Verhulst–Pearl defect model with concurrent failure modes suggests that reliability is likely to be wafer dependent and that changes in threshold saturate for some devices. Both of these are observed in reliability testing and have implications for estimation of activation energy and ultimately lifetime of the devices. Failure modes are manifestations of different defect populations. Different wafers may possess different failure modes that are process dependent. We assume each failure mode is described by a single Verhulst–Pearl defect growth curve and that its rate of growth is characterized by a unique activation energy E_a. Mirroring Fourier analysis, many seemingly different composite degradation curves can be modeled by fitting the curve with multiple independent Verhulst–Pearl failure mode components. It is this remarkable insight that leads to a unified account of degradation curve analysis under a single physical model that we term MCM.

The lifetime estimation of composite degradation curve engenders important rules in conducting ALT, for which a restricted censoring protocol was created. Device life is ultimately governed by the last failure mode, also known as the wear-out failure mode. Following Nash's approach, ALT in the MCM scheme is to be conducted in a time window that is long enough to stabilize all transient failure or annealing modes. Violating this recommendation will undermine the accuracy of the Arrhenius lifetime estimation. Once the transient modes have stabilized, the wear-out mode will be identified and extrapolated to estimate the device lifetime through the Arrhenius relationship.

In section 2.7, we reviewed the methodology in modeling many reported composite degradation curves using MCM and address practical considerations in modeling both linear and nonlinear wear-out failure modes. Previously, all these reported composite degradation curves were described by different device-specific and failure mode–specific empirical models that provide little insight of the common underlying physical process of degradation. Our review articulates how the emergence of MCM unifies our understanding of the underlying degradation processes and sets forth a sensible approach to ALT with physical justifications. It is our hope that this review of MCM will assist in advancing the understanding of device reliability.

2.9. References

Agrawal, G.P., and Dutta, N.K. (1993). *Semiconductor Lasers*, 2nd edition. Van Nostrand Reinhold, New York.

Cassidy, D.T. (2002). Spatially resolved and polarization-resolved photoluminescence for the study of dislocation and strain in III-V materials. *Mater. Sci. Eng. B*, 91–92, 2–9.

Cassidy, D.T., Hall, C.K., Rehioui, O., and Bechou, L. (2014). Strain estimation in III-V materials by analysis of the degree of polarization of luminescence. *Microelectron. Reliab.*, 50, 462–466.

Cassidy, D.T., Lam, S.K.K., Lakshmi, B., and Bruce, D.M. (2004). Strain mapping by measurement of the degree of polarization. *Appl. Opt.*, 43, 1811–1818.

Chang, M.-H., Das, D., Verde, P., and Pecht, M. (2012). Light emitting diodes reliability review. *Microelectron. Reliab.*, 52, 762–782.

Chuang, S.L., Ishibashi, A., Kijima, S., Nakayama, N., Ukita, M., and Taniguchi, S. (1997). Kinetic model for degradation of light-emitting diodes. *IEEE J. Quantum Electron.*, 33, 970–979.

Colbourne, P.D., and Cassidy, D.T. (1993). Imaging of stresses in GaAs diode lasers using polarization-resolved photoluminescence. *IEEE J. Quantum Electron.*, 29, 62–68.

Coldren, L.A., Corzine, S.W., Mashanovitch, M.L. (2012). *Diode lasers and photonic integrated circuits*, 2nd edition. John Wiley & Sons, Hoboken.

DeTemple, T.A., and Herzinger, G.M. (1993). On the semiconductor laser logarithmic gain-current density relation. *IEEE J. Quantum Electron.*, 29, 1246–1252.

Dow, J.D., and Allen, R.E. (1982). Role of dangling bonds and antisite defects in rapid and gradual III-V laser degradation. *Appl. Phys. Lett.*, 41, 672–674.

Dym, C.L., and Ivey, E.S. (1980). *Principles of Mathematical Modeling*. Academic, New York.

Eckler, A.R. (1985). A statistical approach to laser certification. *AT&T Tech. J.*, 64, 765–770.

Escobar, L.A., and Meeker, W.Q. (2006). A review of accelerated test models. *Statist. Sci.*, 21, 552–577.

Fatt, Y.S. (1991). The annealing of double-heterostructure GaInAsP-InP 1.3 μm laser diodes. *IEEE J. Quantum Electron.*, 27(1), 30–39.

Fukuda, M. (1991). *Reliability and Degradation of Semiconductor Lasers and LEDs.* Artech House, Boston.

Fukuda, M., Wakita, K., and Iwane, G. (1983). Dark defects in InGaAsP/InP double heterostructure lasers under accelerated aging. *J. Appl. Phys.*, 54, 1246–1250.

Gause, G.F. (1971). *The Struggle for Existence.* Dover, New York.

Hakki, B.W., Fraley, P.E., and Eltringham, T.F. (1985). 1.3-μm Laser reliability determination for submarine cable systems. *AT&T Tech. J.*, 64, 771–807.

Hartman, R.L., and Dixon, R.W. (1975). Reliability of DH GaAs lasers at elevated temperatures. *Appl. Phys. Lett.*, 26, 239–242.

Hausler, K., Zeimer, U., Sumpf, B., Erbert, G., and Trankle, G. (2008). Degradation model analysis of laser diodes. *J. Mater. Sci.: Mater. Electron.*, 19, 160–164.

Horikoshi, Y., Kobayashi, T., and Furukawa, Y. (1979). Lifetime of InGaAsP-InP and AlGaAs-GaAs DH lasers estimated by the point defect generation model. *Jpn. J. Appl. Phys.*, 18, 2237–2244.

Huang, J.-S. (2013). Burn-in aging behavior and analytical modeling of wavelength-division multiplexing semiconductor lasers: Is the swift burn-in feasible for long-term reliability assurance? *Advances in OptoElectronics*, 24, 1–4.

Jalonen, M., Toivonen, M., Savolainen, P., Kongas, J., and Pessa, M. (1997). Effects of rapid thermal annealing on GaInP/AlGaInP lasers grown by all-solid-source molecular beam epitaxy. *Appl. Phys. Lett.*, 71, 479–481.

Jimenez, J. (2003). Laser diode reliability: Crystal defects and degradation modes. *C. R. Physique*, 4, 663–673.

Joyce, W.B., Liou, K.-Y., Nash, F.R., Bossard, P.R., and Hartman, R.L. (1985). Methodology of accelerated aging. *AT&T Tech. J.*, 64, 717–764.

Kallstenius, T., Landstedt, A., Smith, U., and Granestrand, P. (2000). Role of nonradiative recombination in the degradation of InGaAsP/InP-based bulk lasers. *IEEE. J. Quantum Electron.*, 36, 1312–1322.

Kamejima, T., Ishida, K., and Matsui, J. (1977). Injection-enhanced dislocation glide under uniaxial stress in GaAs-(GaAl)As double heterostructure laser. *Jpn. J. Appl. Phys.*, 30, 233–240.

Kimerling, L.C. (1978). Recombination enhanced defect reactions. *Sol. St. Electron.*, 21, 1391–1401.

Klinger, D.J., Nakada, Y., and Menendez, M.A. (eds). (1990). *AT&T Reliability Manual.* Van Nostrand Reinhold, New York.

Kobayashi, T. and Furukawa, Y. (1979). Recombination enhanced annealing effect in AlGaAs/GaAs remote junction heterostructure lasers. *IEEE J. Quantum Electron.*, 15(8), 674–684.

Lam, S.K.K., Cassidy, D.T., and Mallard, R.E. (2005). Characterization of $SiO_x/Si/SiO_x$ coated n-InP facets of semiconductor lasers using spatially-resolved photoluminescence. *Jpn. J. Appl. Phys.*, 44(11), 8007–8009.

Lam, S.K.K., Mallard, R.E., and Cassidy, D.T. (2003a). Analytical model for saturable aging in semiconductor lasers. *J. Appl. Phys.*, 94, 1803–1809.

Lam, S.K.K., Mallard, R.E., and Cassidy, D.T. (2003b). Effects of having two populations of defects growing in the cavity of a semiconductor laser. *J. Appl. Phys.*, 94, 2155–2161.

Lam, S.K.K., Mallard, R.E., and Cassidy, D.T. (2004). An extended multi-component model for the change of threshold current of semiconductor lasers as a function of time under the influence of defect annealing. *J. Appl. Phys.*, 95, 2264–2271.

Lewis, E.E. (1996). *Introduction to Reliability Engineering*, 2nd edition. John Wiley & Sons, Hoboken.

Lisak, D., Cassidy, D.T., and Moore, A.H. (2001). Bonding stress and reliability of high power GaAs-based lasers. *IEEE Trans. Comp. Packag. Technol.*, 24, 92–98.

Maki, D.P., and Thompson, M. (1973). *Mathematical Models and Applications.* Prentice, Hoboken.

Nash, F.R. (1993). *Estimating Device Reliability: Assessment of Credibility.* Kluwer Academic Publishers, Boston.

Nash, F.R., Joyce, W.B., Hartman, R.L., Gordon, E.I., and Dixon, R.W. (1985). Selection of a laser reliability assurance strategy for a long-life application. *AT&T Tech. J.*, 64, 671–715.

Nelson, W. (1990). *Accelerated Testing: Statistical Models, Test Plans, and Data Analyses.* John Wiley & Sons, New York.

Newman, D., and Ritchie, S. (1973). Gradual degradation of GaAs double-heterostructure lasers. *IEEE J. Quantum Electron.*, 9(2), 300–305.

O'Hara, S., Hutchinson, P.W., and Dobson, P.S. (1977). The origin of dislocation climb during laser operation. *Appl. Phys. Lett.*, 30, 368–370.

Ohring, M. (1998). *Reliability and Failure of Electronic Materials and Devices.* Academic Press, San Diego.

Okayasu, M., and Fukuda, M. (1992). Estimation of the reliability of 0.98 μm InGaAs/GaAs strained quantum well lasers. *J. Appl. Phys.*, 72, 2119–2124.

Pearl, R., and Reed, L.J. (1920). On the rate of growth of the population of the united states since 1790 and its mathematical representation. *Proc. Nat. Acad. Sci. USA*, 6, 275–288.

Petroff, P., and Hartman, R.L. (1973). Defect structure introduced during operation of heterojunction GaAs lasers. *Appl. Phys. Lett.*, 23, 469–471.

Petroff, P.M., and Kimerling, L.C. (1976). Dislocation climb model in compound semiconductors with zinc blende structure. *Appl. Phys. Lett.*, 29, 461–463.

Sim, S.P. (1990). A review of the reliability of III-V opto-electronic components. In *Semiconductor Device Reliability*, Christou, A., and Unger, B.A. (eds). Kluwer Academics, Dordrecht, 301–319.

Telcordia Technologies (1998). Generic Reliability Assurance Requirements for Optoelectronic Devices Used In Telecommunications Equipment, GR-468-CORE, Standard, Issue 1, 1998.

Tobias, P.A., and Trindade, D.C. (1995). *Applied Reliability*, 2nd edition. Van Nostrand Reinhold, New York.

Ueda, O. (1996). *Reliability and Degradation of III-V Optical Devices*. Artech House, Boston.

Ueda, O. (2010). On degradation studies of III-V compound semiconductor optical devices over three decades: Focusing on gradual degradation. *Jpn. J. Appl. Phys.*, 49, 090001:1–8.

Ueda, O., Isozumi, S., Kotani, T., and Yamaoka, T. (1977). Defect structure of $<100>$ dark lines in the active region of a rapidly degraded $Ga_{1-x}Al_xAs$ LED. *J. Appl. Phys.*, 48, 3950–3952.

Verhulst, P.F. (1838). Notice sur la loi que la population suit dans son accroissement. *Corr. Math. et Phys. publ. par A. Quetelet*, T.X., 113–121.

Werner, J., Kapon, E., Lehmen, A.C.V., Bhat, R., Colas, E., Stoffel, N.G., and Schwarz, S.A. (1988). Reduced optical waveguide losses of a partially disordered GaAs/AlGaAs single quantum well laser structure for photonic integrated circuits. *Appl. Phys. Lett.*, 53, 1693–1695.

Yamada, N., Roos, G., and Harris, J.S. (1991). Threshold reduction in strained InGaAs single quantum well lasers by rapid thermal annealing. *Appl. Phys. Lett.*, 59, 1040–1042.

Zhang, G., Nappi, J., Ovtchinnikov, A., Asonen, H., and Pessa, M. (1992). Effects of rapid thermal annealing on lasing properties of InGaAs/GaAs/GaInP quantum well lasers. *J. Appl. Phys.*, 72, 3788–3791.

3

Reliability of Laser Diodes for High-rate Optical Communications – A Monte Carlo-based Method to Predict Lifetime Distributions and Failure Rates in Operating Conditions

3.1. Introduction

High performances and high reliability are two of the most important goals driving the penetration of optical transmission into telecommunication systems ranging from 880 to 1550 nm. However, performance of high-rate optical communications systems is strongly related to micro-optoelectronic device parameters inserted into the transmitter and receiver blocks that have been used in telecommunication applications. These blocks contain InP photonic and GaAs or InP electron devices. It has already been demonstrated that hybrid or monolithic integration used to assemble these modules (optoelectronic integrated circuits (OEICs)) suffer from various intrinsic and process-dependent parasitic effects. In this case, reliability estimation is traditionally based on life-testing and a current approach is to apply Telcordia requirements (468GR) for optoelectronic applications (Park *et al.* 2003) and lifetime prediction, defined as the time at which a parameter reaches its maximum acceptable shift, is the main outcome in terms of reliability estimation for a technology. For optoelectronic emissive components, selection tests and life testing are specifically used for reliability assessment (in particular

Chapter written by Laurent Mendizabal, Frédéric Verdier, Yannick Deshayes, Yves Ousten, Yves Danto and Laurent Béchou.

long-term reliability prediction in operating conditions for a given mission profile) according to well-established qualification standards (i.e. Telcordia GR-468 CORE). This approach is based on an extrapolation of degradation laws from the time-dependent drift of electrical or optical parameters and associated physics of failure, allowing strong test time reduction.

The main problem is that such aging tests must be performed on more than a few hundreds (at least) of devices aged over five to ten thousand hours mixing different temperatures and drive current conditions conducting to acceleration factors above a few hundreds (>100). These conditions are costly, time-consuming and cannot give a complete distribution of times to failure (Béchou 2015). Moreover, the actual trend for qualification procedures is to reduce the number of components for qualification tests.

In such a context, actual levels of reliability regarding extended failure times and very low failure rates lead to a dramatic increase in difficulty for experimental evaluation. Moreover, the complexities of systems using infrared optics, used to align component and fiber, increase the difficulty in localizing degraded zones responsible for optical power loss. Actual performances, required by the end-user, force the manufacturer to perfectly predict the long-term behavior of each device constituting the system (Svensson 2004). However, due to the various interactions between devices, the reliability of the system cannot be established as the reliability of the less reliable part (Wang 2004).

In a previous paper, an original methodology was presented to estimate the long-term performance degradation of optical communication links based on the wavelength-division multiplexing (WDM) technique, by using a dedicated system simulator for predictive qualification and design for reliability. In particular, we have presented two frameworks of point-to-point optical communication systems in the space environment (i.e. satellites) where high-rate digital data (or frequency bandwidth), lower cost or mass saving are needed (Béchou 2013). Optoelectronics devices used for these applications can be similar to those found in systems for terrestrial optical networks. Particularly, we have reported simulation results of transmission performances after the introduction of time-dependent variations (assuming a 'power' law $\sim at^m$) of distributed feedback (DFB) Laser diode functional parameters extrapolated from accelerated tests mixing temperature (80°C) and constant injected current (150 mA).

Simulations using the COMSIS© simulator were performed to investigate and predict the consequence of DFB transmitter degradations (acting as an optical frequency carrier) on key indicators of network performance (i.e. eye diagram, noise, cross-talk between channels, quality factor, Q-factor, and bit error rate, BER). The studied link consisted in 4 channels, each operating at 2.5 Gbits/s based on the WDM technique with direct modulation. The transmitters are equally spaced by

100 GHz (0.8 nm) around the 1550 nm center wavelength. The results clearly demonstrated that variation of functional parameters, such as modulated bias current and central wavelength, induces a penalization in dynamic performances of the complete WDM link. For a statistical representativeness, different degradation kinetics of aged laser diodes from a same batch have been implemented to build the final distribution of Q-factor and BER values after 25 years. The strong interest of such system simulations is presented in order to highlight the impact of component parameter degradations on the whole network performances, carry out various sensitivity analyses of the product development. Thus the validity of failure criteria in relation with operating conditions could be evaluated representing a significant part of the general PDfR (probabilistic design for reliability) effort (Suhir 2012). Nevertheless, even if we have shown the strong relevance to introduce experimental parametric distributions after accelerated tests, rather than consider the atypical behavior of a single laser diode to characterize the impact of such degradation on high-rate optical network performances, such an approach remains very challenging and time-consuming in order to be performed in a qualification procedure at the industrial level in particular because of the increasingly reduced time-to-market.

Concerning laser diode reliability, the main problem is that actual components, used in the optical transceiver system, demonstrate extremely low failure rates and the determination of lifetime distribution is more difficult. For instance, distributed feedback single mode laser diodes (DFB-LD) used as optical carrier for 1550 nm telecommunication networks present median lifetime close to 10^6 h at room temperature (25°C) and constant optical power around 10 mW.

Figure 3.1 clearly shows that, for operating conditions, the time to observe the first failures (i.e. the first failed device) with 90% confidence level α is higher than 10^7 h. In order to reduce this time, it is necessary to accelerate the failure process by aging conditions associated with a large increase in temperature (T_{acc}). For such a technology, it is possible to apply aging conditions by increasing the temperature until it reaches 100°C, allowing us to analyze drift of monitored static electro-optical parameters such as bias current, output optical power, and center wavelength (Bonfiglio 1998). Considering such an increase in temperature, it is possible to ensure that the activation energy (E_a) is constant. It is assumed that the same type of failures will be observed for aging but also for operating conditions guaranteeing the calculated lifetime from the experimental aging test results. This lifetime assessment is based on failure criteria set for this technology, in particular with respect to the drift of central wavelength which must be lower than 0.1 nm at 1550 nm at 100 mA. But, actual experimental results show that even after a long aging test (close to 10^4 h), the failure criteria is not reached and sometimes only 5% drift of monitored parameters is observed predicting "extremely" low failure rate (Mendizabal 2006). The typical general failure rate function λ(t) of DFB 1550 nm laser diode is presented in Figure 3.2, which can be separated into three different zones:

– the failure process is responsible for rapid or catastrophic optical power losses observed during burn in tests. The roller-coaster curve, shown in the first part of the λ(t) variation as plotted in Figure 3.2, is related to a small number of failure mechanisms induced by the manufacturing process;

– the operating lifetime is characterized by a low failure rate that is difficult or impossible to predict. This part of the failure rate curve underlies the actual problem for lifetime distribution estimation, but we are going to demonstrate that the Monte Carlo method will be a possible issue as already described in many examples targeting laser diodes applications for the most part (Bao 2011; Zio 2015; Liu 2017; Maliakal 2019);

– the last part of the curve represented the wear phenomenon related to extrinsic defect induced by the mission profile, as defined by the end-user in telecommunication applications.

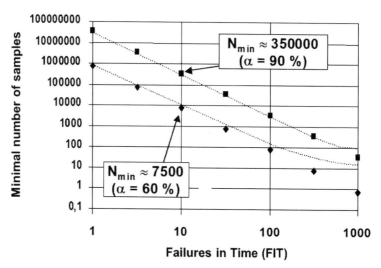

Figure 3.1. *Minimal number of samples versus the number of FIT to obtain 50% of failures for two upper confidence levels α (in this example, an acceleration factor of 350 is assumed)[1]*

1 Acceleration factor (AF): the acceleration factor is given when a kind of stress similar to use stress applied to the device in a very short period of time results in device failure (Bernstein 2014).

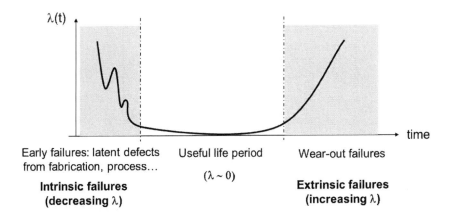

Figure 3.2. *Variations of the time-dependent failure rate (so-called "roller-coaster" curve) for a modern and mature electronic device*

Facing actual qualification challenges, a new qualification approach is proposed by focusing on reliability concerns at the early stage of product development (Imai *et al.* 1978). This method has already proven to be of great importance and the results should help to implement component design rules based on physics of failure knowledge. One of the first activities is to simulate long-term drift of electro-optical parameters to build a decay model of the component. From this model, the interest of this methodology is to build virtual components allowing access to the first times of failure for the technology studied. This statistic method, based on the Monte Carlo approach, allows us to extrapolate the average lifetime and number of FIT (failure in time)[2] from accelerated and short-time tests.

3.2. Methodology description

3.2.1. *Application context*

The method presented in this section is based on a combination of techniques for estimating and predicting lifetime, in the sense that the statistical study carried out is based on experimental data obtained from the accelerated aging of a batch of components. These accelerated aging tests can be performed according to different test conditions:

– passive storage (temperature only);

– active storage (with component supply).

---

2 Failure in time (FIT): 1 FIT corresponds to a failure on one billion devices (109) per hour.

However, considering the very long lifetime of modern technologies and devices, the purpose of these tests is no longer to observe the time taken by a parameter to exceed a chosen failure criterion with respect to operating conditions. It is necessary to focus on the variation kinetics of functional parameters during a given time period, generally less than 5000 h.

The results of these tests are then used by fitting the curves obtained from variations of functional parameters using a mathematical degradation law. Such a drift can be empirical when it is only related to the mathematical smoothing of these curves, or physical when based on the modeling of a physical mechanism related to the device. An example of a well-established empirical law is illustrated through the so-called "power" law (at^m).

The latter has been extensively used to model the degradation kinetics of a large number of devices, for example NMOS transistors with the observation of the threshold voltage (V_{th}) as a function of time (Nougier 1987; Cui 2005), metal/insulator/metal (MIM) capacitors in which the lifetime and the electric field are linked (Hwang *et al.* 1997; Bolam 2002), or optoelectronic devices such as photodiodes (dark current) (Sauvage *et al.* 2000; Telcordia GR-468-CORE 1998; Kim 2004). This law is also commonly used to model the degradation kinetics of the operating bias current of ultra-stable transceivers for optical links (Park *et al.* 2003; Ikegami and Fukuda 1991; Sim 1993; Mawatari 1996). An example of this adjustment is presented in Figure 3.3, in the case of the threshold current (I_{th}) variations of a 1550 nm DFB laser diode.

Once the adjustment has been made using a degradation law for all the devices considered for the test, we obtain a number N of couples (a, m) with N generally much less than 100. These N couples represent all the devices of our study, from the point of view of the degradation kinetics. However, this number of N components are too small to be statistically representative, and new couples (a, m) must be created in order to reinforce the statistical credibility necessary within the framework of these approaches. These couples, which we will name "virtual couples", are created by a random drawing according to the Monte Carlo method, which we will detail in the following section.

The choice of the number N of elements constituting the sample is left to the user, who must take into account the compromise between the number of elements and the computation time; the latter is not negligible when the number of drawings becomes high.

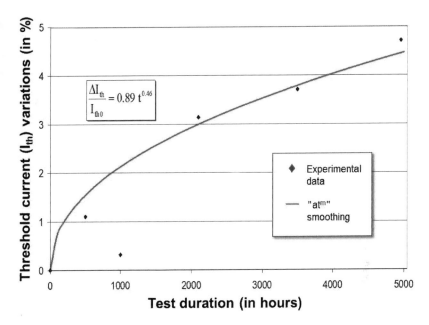

Figure 3.3. *Smoothing of threshold current (I_{th}) variations using a power law (at^m), measured on a DFB 1550 nm laser diode during an aging test (active storage)*

The next step is to calculate, from the couples (a, m) previously obtained, the lifetime corresponding to each of them. For this, an assumption must be made, according to which the extrapolation of the parametric variation curves – until the failure criterion is reached – is possible from the same degradation law; in other words, the degradation mechanism undergone by the device studied is only described by a single degradation law throughout the duration of its aging, extracted from experimental data. Indeed, it is quite possible that a degradation mechanism, during its formation, leads to additional degradations with its own kinetics and a specific failure signature. By considering this assumption, the lifetime (t_{eol}) corresponding to each couple is calculated from a failure criterion δ, defined as follows:

$$t_{eol} = \sqrt[m]{\frac{\delta}{a}} \qquad\qquad [3.1]$$

Finally, from the obtained distribution of lifetimes, we can then deduce the associated failure rate λ.

3.2.2. *Monte Carlo random sampling*

Monte Carlo methods are known and derived from applications of experimental mathematics focused on random numbers. Many applications can be addressed and in particular:

– the field of nuclear physics;

– chemistry;

– biology;

– medicine.

There are two types of problems that can occur: probabilistic or deterministic depending on whether or not they directly concern the behavior and the result of random processes.

In the case of a probabilistic problem, the Monte Carlo approach in its simplest form consists of directly observing chosen random numbers in such a way that they simulate the random physical process to be studied, and in deducing the desired solution of the behavior of these random numbers. A well-established example given by J. M. Hammersley corresponds to the study of the growth of an insect population based on demographic data of survival and reproduction (Hammersley 1964). We can thus assign, to each of the individuals, random numbers representative of their age at the birth of these descendants then another at their death, thereafter processing their descendants and the following generations in the same way. By making these numbers agree with the demographic statistics, one can obtain a random sample of the population that we can use as if it were data collected by the entomologist in the laboratory or in the field. But these artificial data points have the advantage of being easier to collect, or they leave to demographic statistics a latitude of variation proscribed by nature. In this case, one can refer to a direct Monte Carlo simulation.

Now suppose that we are faced with a deterministic problem that we can formulate in theoretical language but not solve by theoretical means. Being deterministic, this problem has no direct relationship with random processes, but once the theory has exposed its underlying structure, we can *a priori* recognize that such a structure or this formal expression also describes some random process, apparently unrelated. We can thus solve the deterministic problem numerically by a Monte Carlo simulation of the corresponding probabilistic problem. For example, a theoretical problem of electromagnetism may require the resolution of a Laplace equation subject to boundary conditions that defeat the theory of classical analysis. However, this equation can be used in particular to describe the movement of particles diffusing in a random way in a field limited by absorbent barriers. The electromagnetic problem can be solved by performing an experiment where the particles could be guided by means of random numbers, defining the parameters of the guide and/or particles until they are absorbed on barriers specially chosen to represent these boundary conditions.

This technique given by a Monte Carlo simulation of a different problem can be called the "elaborate" Monte Carlo method to distinguish them from the direct simulation of the initial problem.

In our particular case, the performed simulation will consist of a draw of two independent distributed random variables, corresponding to the direct Monte Carlo method. However, before going into the details of this method, we give hereafter a brief review of the main elements relating to random numbers.

3.2.3. *Brief review on random and pseudo-random numbers*

From a practical way, two solutions are used to obtain random numbers (Nougier 1987):

– pseudo-random number tables;

– calculator realizing this draw under specific conditions.

The use of a calculator leads to the construction of pseudo-random numbers. These numbers are different from the random numbers – when the first element(s) of a series of pseudo-random numbers are chosen arbitrarily, the whole sequence is completely determined: the sequence is then reproducible, which makes it possible to verify the programs in which they are used. In addition, since their properties are identical to those of random numbers, it allows the substitution of a series of pseudo-random numbers by a series randomly obtained, and this in most calculations.

A basic equation enables us to generate a series of pseudo-random numbers with uniform distribution as follows:

$$x_i = a.x_{i-1} + c \text{ (modulo m)} \qquad [3.2]$$

This means that the number x_i is equal to the remainder of the division by m of the second member of the equation. The quantities m, a and c are constants.

The obtained numbers have a uniform distribution (also called rectangular). Most computers give sequences of pseudo-random numbers with rectangular distribution over [0,1]. These numbers, called r_i, are obtained from the x_i generated values by

[3.2] by noting that the remainder of a division is less than the divisor. As a result, we get:

$$r_i = \frac{x_i}{m-1}$$ [3.3]

Similarly, when we need to generate numbers with uniform distribution in the interval [a, b], we associate a number ξ with any random r value uniformly distributed on [0,1] and defined by:

$$\xi = a + (b - a).r$$ [3.4]

3.2.4. Direct Monte Carlo method: non-rectangular distribution

The determination of distribution functions requires the generation of random numbers (ξ) whose distribution is not uniform over [a, b], and takes as values f(x). This means that if we generate this set of random numbers (ξ), the proportion of these numbers such that $x \leq \xi \leq x + dx$ is then equivalent to f(x)dx. An equivalent way of expressing the latter is to say that the probability to have ξ is included into the interval [x, x + dx] equals to f(x)dx.

Let F(x) be the distribution function of f(x), defined by:

$$F(x) = \int_a^x f(u)\,du$$ [3.5]

For $a \leq x \leq b$, we obtain $0 \leq F(x) \leq 1$. Given a random number r with uniform distribution on [0,1], the relationship between r and x is given as:

$$r = F(\xi) = \int_a^\xi f(u)\,du$$ [3.6]

The distribution function of x, i.e. g(x), is such that:

$$dP = \text{Prob}\left\{ x \leq \xi \leq x + dx \right\} = g(x)\,dx$$ [3.7]

Note that the probability dP that ξ is in the interval dx corresponds to the probability that r is in the interval dF, and we obtain:

$$dP = dF = \frac{dF}{dx}dx = f(x)\,dx$$ [3.8]

By comparing equations [3.7] and [3.8], we see that f(x) = g(x). Consequently, the random variable ξ is therefore described by the distribution f(x).

This technique is the most direct way to generate random variables from a distribution. It can be applied whenever equation [3.6] has an analytical solution or can be solved with respect to the random variable ξ.

In order to obtain the distribution f(x), we must first generate a large number N of random numbers r and deduce the associated random numbers ξ. Then, by counting the values $\Delta n(x)$ of x located in the interval defined by $\left[x - \dfrac{\Delta x}{2}, x + \dfrac{\Delta x}{2} \right]$, we have:

$$\frac{\Delta n(x)}{N} = f(x)\, \Delta x \qquad\qquad [3.9]$$

We can thus construct the distribution histogram. The precision obtained is then correlated to the fact that N and Δx must be, respectively, large and small.

3.3. Description of the experimental approach

3.3.1. *Choice of the correlation law*

The Monte Carlo method is used for drawing couples (a, m), where a and m represent the parameters associated with the degradation law of a parameter P_a such that:

$$\frac{\Delta P_a}{P_a} \propto at^m \qquad\qquad [3.10]$$

The generation of distribution laws from virtual couples (a, m) makes it possible to obtain a joint distribution law denoted as $F_{a,m}$. The choice of each distribution law is left to the user, based on the following three main points:

– normal law;

– log-normal law;

– the so-called Weibull's law.

The parameterization of these laws is carried out from the distribution laws of the parameters (a, m), obtained from experimental data.

Observation shows that the coordinates of the couples (a, m) are not independent. The search for a joint distribution function $F_{a,m}$ is, in this case, without a single solution. The determination of a correlation trend law between a and m here makes it possible to obtain the joint distribution by successive determination of the marginal laws associated, on the one hand, with a coordinate declared arbitrarily "principal", and, on the other hand, with the deviation of the other coordinate from the previous trend law.

To illustrate our point, let us take a concrete example. We choose a correlation law between a and m of logarithmic type such that:

$$g(a) = A \ln(a) + B$$
$$\text{A and B are constants}$$

[3.11]

Δm will then be the difference between the experimental coordinate m and g(a):

$$\Delta m = m - A \ln(a) - B$$

[3.12]

If the trend law is optimized, this difference is comparable to noise on the second coordinate, and therefore independent of the first. A homogeneous couple in the starting lot can then be reconstituted by drawing the main coordinate, followed by drawing a deviation from the secondary coordinate, superimposed on the trend value obtained from the first coordinate.

The choice of the logarithmic correlation must be relevant. Indeed, this correlation by taking **m** as a secondary variable seems the most suitable in our case. The approach of this interpolation phase is shown in Figure 3.4.

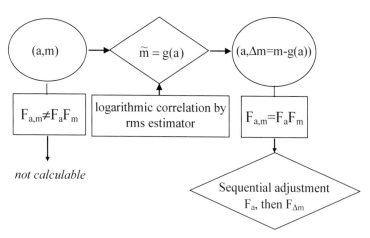

Figure 3.4. *Proposed approach for an interpolation process (rms: root mean square)*

3.3.2. *Calculation of failure times and failure rates*

After having drawn the N couples (a, Δ m), the construction of the couples (a, m) is carried out by a reconstruction of m considering equation [3.12]. Then, for each couple, the end-of-life time t_{eol} is calculated from equation [3.1] and a predefined failure criterion.

These N instants of failure will be adjusted using a log-normal or Weibull-type law in order to calculate their probability density $f(t_{eol})$. From the latter, the distribution law of the lifetimes of the devices $F(t_{eol})$ is then calculated, i.e. the distribution function of failures over time.

From this distribution, one can then calculate $\lambda(t)$, the instantaneous failure rate defined as the derivative with respect to time t of the distribution of the failed devices, divided by the number of components having survived at time t. Thus:

$$\lambda(t) = \frac{dF(t)}{dt} \times \frac{1}{1-F(t)} \qquad [3.13]$$

The user can then either plot the failure rate over time or calculate the number of FITs corresponding to a mission profile (i.e. operating conditions of the device). This process is shown schematically in Figure 3.5, highlighting all the steps proposed in the context of an extrapolation.

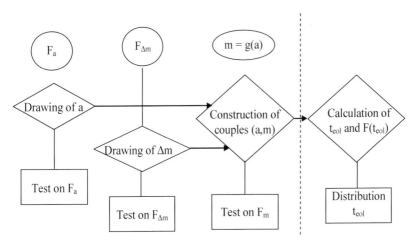

Figure 3.5. *Proposed approach for an extrapolation process*

3.3.3. *Application to DFB laser diodes emitting at 1550 nm*

In this part, we will endeavor to demonstrate the feasibility of the proposed methodology, through an example, with the objective of starting from the extraction of the couples (a, m) enabling the estimation of the associated failure rate of the studied components. In the next section, the results of this feasibility demonstration will be validated by an analytical approach taking into account the original couples (a, m). This demonstration was carried out from the drifts of the I_{Bias} supply current on a batch of 30 DFB 1550 nm laser diodes. These 30 diodes were distributed over three different test vehicles (TV) and aged through active storage conditions (53°C, 110 mA) corresponding to accelerated conditions and during over 5000 hours. Among these three TVs, 8 components were chosen so that different types of variation are represented. Indeed, a certain consistency was observed between the behaviors of the devices of the same TV, which would be likely to bias the feasibility study. The couples (a, m) extracted, from an adjustment of the supply current from a power function, are presented in Table 3.1.

Chip No.	a	m
1	0.011	0.59
2	0.1	0.4
3	0.5	0.3
4	0.18	0.4
5	0.2	0.35
6	0.18	0.3
7	0.02	0.5
8	0.025	0.4

Table 3.1. *Experimental extrapolated numbers (a, m) obtained from a batch of 8 aged DFB laser diodes emitting at 1550 nm (active storage conditions: 53 °C, 110 mA, 5000 h)*

The distribution of the couples and the associated correlation law are jointly represented in Figure 3.6. The optimal logarithmic correlation law was obtained by the least-squares method.

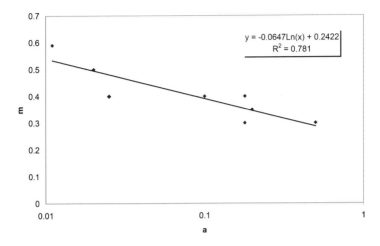

Figure 3.6. *Correlation law of experimental couples (a, m) distribution obtained from a simple ordinary least-squares method*

From this correlation law, it is possible to calculate an approximation of the distributions of the main variable **a** and the secondary variable **Δm**, residue of the comparison between the experimental value **m** and that obtained from the correlation law. The distribution law of the main variable was chosen as being represented by a log-normal law (equation [3.14]), while the law associated with the secondary variable corresponds to a normal law (equation [3.15]), more representative experimental noise. The parameters of each of these laws obtained during the approximation calculation are summarized in Table 3.2.

$$G\left(m_x, s_x\right) = \frac{1}{2}\left[1 + \mathrm{erf}\left(\frac{\ln x - m_x}{\sqrt{2}s_x}\right)\right] \tag{3.14}$$

$$N\left(\mu_x, \sigma_x\right) = \frac{1}{2}\left[1 + \mathrm{erf}\left(\frac{x - \mu_x}{\sqrt{2}\sigma_x}\right)\right] \tag{3.15}$$

Main variable: a	Log-normal law	Median: $m_x = 0.061$
		Standard deviation: $s_x = 1.679$
Secondary variable: Δm	Normal law	Average: $\mu_x = 0.061$
		Standard deviation: $\sigma_x = 0.045$

Table 3.2. *Associated parameters to the distribution laws of the main and secondary variables*

Finally, Figure 3.7 presents the results obtained for a draw of 1024 virtual couples: Figure 3.7(a) gives a comparison of the joint distributions $F_{a,m}$ of the experimental and simulated couples (a,m), and Figure 3.7(b) plots the distribution of the error between these two cumulative distributions. This error, less than 0.4%, demonstrates a good correlation between the distributions from experimental and simulated couples. Figure 3.7(c) allows a comparison of the distributions of experimental and simulated couples (a, m).

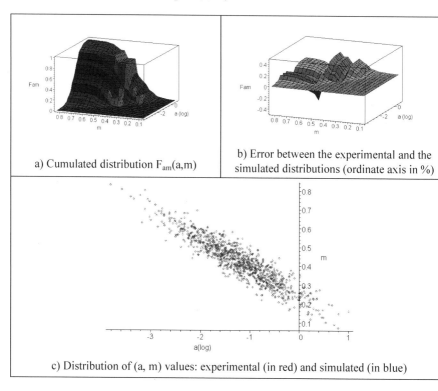

a) Cumulated distribution $F_{am}(a,m)$

b) Error between the experimental and the simulated distributions (ordinate axis in %)

c) Distribution of (a, m) values: experimental (in red) and simulated (in blue)

Figure 3.7. *Results of random sampling using 1024 sets of virtual couples (a, m) and comparison with experimental couples (a, m). For a color version of this figure, see www.iste.co.uk/vanzi/reliability.zip*

Once the new couples (a, m) have been created, the equation recalled below is applied to each of them in order to construct the distribution of the instants of failure presented in Figure 3.8(a). This distribution is then approached by a log-normal law (see Figure 3.8(b)) in order to calculate the instantaneous failure rate $\lambda(t)$ (Figure 3.8(c)). Finally, this instantaneous failure rate is calculated for a fixed lifetime of the device, also indicating the number of associated FITs.

The z coordinate in Figure 3.8(b) corresponds to the following change of variable (see equation [3.16]), making it possible to fit a line called "Henry's line", estimating the validity of a distribution to fit normal behavior (here, log-normal).

$$z = \left(\frac{1}{2} + \frac{1}{2} \text{erf} \right)^{-1} (x) \tag{3.16}$$

In our case, the failure rate was calculated at 20 years. However, since the result obtained is dependent on the random distribution of the points along the chosen distribution law, several prints were necessary in order to limit this error, which is difficult to quantify analytically. The results are compared in Table 3.3 for a number of pairs (a, m) drawn from 100 and 1024 up to 5 successive random draws.

Number of simulated points	Adjusted failure rate for accelerated aging conditions (*and for operating conditions*) in FITs at 20 years						
	First sampling	Second sampling	Third sampling	Fourth sampling	Fifth sampling	Average	Standard deviation
100	1051 (131)	1018 (127)	697 (87)	1107 (138)	1009 (126)	980 (122)	160 (20)
1024	944 (118)	984 (123)	908 (113)	906 (113)	924 (115)	930 (120)	30 (8)

Table 3.3. *Failure rate obtained for two sampling processes (based on 100 and 1024 points) in accelerated conditions (acceleration factor – AF = 8) and in operating conditions*

One can see the impact of the number of pairs (a, m) drawn on the failure rate, which is mainly at the standard deviation of the different draws, rather than at the level of the average value of the results of successive draws. For a draw of 100 points, the average failure rate is 980 FITs, with a standard deviation of 160, while for 1024 points the average is 930 FITs and the standard deviation is 30.

A compromise therefore remains to be determined between the accuracy of the result, in the sense of repeatability, and the time allocated by the user for such an operation. We need to keep in mind that this non-repeatability of the result is inevitable and inherent in the method.

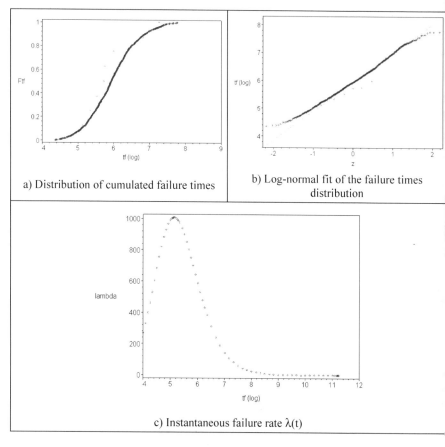

a) Distribution of cumulated failure times

b) Log-normal fit of the failure times distribution

c) Instantaneous failure rate λ(t)

Figure 3.8. *Distribution of cumulated failure times and instantaneous failure rate. For a color version of this figure, see www.iste.co.uk/vanzi/reliability.zip*

3.4. Robustness analysis of the proposed method

3.4.1. *Validation of the method: analytical approach*

3.4.1.1. *Brief theoretical background*

The purpose of this section is to show the influence of the bias, if it exists, that can be committed by the use of the Monte Carlo method in our methodology. For this, we replace the part corresponding to the drawing of the random variables, noted here A and ΔM, having values a and Δm, respectively, by an analytical computation based on the distributions, and determined experimentally.

First, we will therefore present the distribution laws to be chosen by the user. There are three laws: the normal law, the log-normal law and the Weibull's law. We will recall, for each one, the probability density as well as the distribution law in Table 3.4.

	Probability density laws	Distribution laws
Normal law	$n_X(x) = \dfrac{1}{\sigma\sqrt{2\pi}} e^{-\frac{(x-x_0)^2}{2\sigma^2}}$	$N_X(x) = \dfrac{1}{2}\left[1+\operatorname{erf}\left(\dfrac{x-x_0}{\sigma\sqrt{2}}\right)\right]$
Log-normal law	$g_Y(y) = \dfrac{1}{s\sqrt{2\pi}}\dfrac{1}{y} e^{-\frac{(\ln y - \ln y_0)^2}{2s^2}}$	$G_Y(y) = \dfrac{1}{2}\left[1+\operatorname{erf}\left(\dfrac{\ln y - \ln y_0}{s\sqrt{2}}\right)\right]$
Weibull's law	$w_Z(z) = \dfrac{\beta}{\alpha}\left(\dfrac{z}{\alpha}\right)^{\beta-1} e^{-\left(\frac{z}{\alpha}\right)^{\beta}}$	$W_Z(z) = 1 - e^{-\left(\frac{z}{\alpha}\right)^{\beta}}$

Table 3.4. *Probability density and distribution laws considered in our study*

In our case, and for the sake of consistency, we will choose the parametric distribution laws used in the previous example. Thus, the parameters a and Δm, defined by equation [3.17], and serving as a starting postulate for our calculation, will therefore be distributed taking into account the following conditions:

– a log-normal distribution with probability density $g_A(a)$ with parameters l_a and s;

– and a normal law of probability density $n_{\Delta M}(\Delta m)$ with zero mean value and standard deviation σ.

$$m = \alpha \ln(a) + \beta + \Delta m \qquad [3.17]$$

Since A and ΔM are two independent random variables, we can write the joint probability density of the random variables A and ΔM in the form:

$$f_{A,\Delta M}(a, \Delta m) = g_A(a) \times n_{\Delta M}(\Delta m) \qquad [3.18]$$

We will determine the joint probability density of the random variables A and M, which we will call $f_{A,M}(a,m)$. By definition, we have:

$$\begin{pmatrix} a \\ \Delta m \end{pmatrix} \mapsto \begin{cases} a = a \\ m = \alpha \ln(a) + \beta + \Delta m \end{cases} \qquad [3.19]$$

However, this system can be written as:

$$\begin{cases} a = a \\ m = \alpha \ln(a) + \beta + \Delta m \end{cases} \Leftrightarrow \begin{cases} a = a \\ \Delta m = m - \alpha \ln(a) - \beta \end{cases} \qquad [3.20]$$

Let us use this last system for a variable change and by calculating the inverse Jacobian operator given by:

$$J^{-1} = \begin{vmatrix} 1 & 0 \\ -\dfrac{\alpha}{a} & 1 \end{vmatrix} = 1 \qquad [3.21]$$

We can then write:

$$f_{A,M}(a,m) = g_A(a) \times n_M(m - \alpha \ln a - \beta) \qquad [3.22]$$

Noting that:

$$n_M(m - \alpha \ln(a) - \beta) = \frac{1}{\sigma\sqrt{2\pi}} e^{-\frac{(m - \alpha \ln(a) - \beta)^2}{2\sigma^2}} = \frac{1}{\sigma\sqrt{2\pi}} e^{-\frac{(\ln(a^\alpha) - m + \beta)^2}{2\sigma^2}} \qquad [3.23]$$

We can recognize the form of a log-normal probability density that would take as random variable A^α, and as parameters $(\beta - m, \sigma)$, within $1/a^\alpha$.

We obtain:

$$n_M(m - \alpha \ln a - \beta) = a^\alpha g_{A^\alpha}(a^\alpha) \qquad [3.24]$$

We can then write the joint probability density $f_{A,M}(a,m)$ only as a function of the random variable A in the form:

$$f_{A,M}(a,m) = a^\alpha \times g_A(a) \times g_{A^\alpha}(a^\alpha) \qquad [3.25]$$

The marginal probabilities of each of the two variables A and M are then written, respectively:

$$\begin{cases} f_A(a) = \int\limits_{-\infty}^{+\infty} f_{A,M}(a,m)dm \\[4mm] f_M(m) = \int\limits_{-\infty}^{+\infty} f_{A,M}(a,m)da \end{cases}$$ [3.26]

thus:

$$\begin{cases} f_A(a) = \dfrac{1}{\sqrt{2\pi}} \dfrac{e^{-\frac{(\ln(a)-\ln(l_a))}{2s^2}}}{as} = g_A(a) \\[6mm] f_M(m) = \dfrac{1}{\sqrt{2\pi}} \dfrac{e^{-\frac{(m-\alpha\ln(l_a)-\beta)^2}{2\left(\sqrt{\sigma^2+\alpha^2 s^2}\right)^2}}}{\sqrt{\sigma^2+\alpha^2 s^2}} = n_M(m) \end{cases}$$ [3.27]

with $n_M(m)$ taking parameters ($\alpha\ln(l_a)+\beta$, $\sqrt{\sigma^2+\alpha^2 s^2}$).

It is important to note that the parameters relating to marginal probability densities, explained in [3.27], will be assigned a *tilde* unlike the parameters of probability densities $g_A(a)$ and $n_{\Delta M}(\Delta m)$ presented above. We thus get:

$$\begin{cases} \tilde{l}_a = l_a, \ \tilde{s} = s \text{ on } f_A(a) \\ \tilde{\mu}_m = \alpha\ln(l_a)+\beta, \ \tilde{\sigma}_m^2 = \sigma^2+\alpha^2 s^2 \text{ on } f_M(m) \end{cases}$$ [3.28]

We then have the following system of equations:

$$\begin{cases} l_a = \tilde{l}_a \\ s = \tilde{s} \\ \sigma^2 = \tilde{\sigma}_m^2 - \alpha^2\tilde{s}^2 \\ \alpha\ln(\tilde{l}_a)+\beta = \tilde{\mu}_m \end{cases}$$ [3.29]

However, parameters l_a and s are fixed by the probability density of parameter a, itself calculated from experimental data. In the same way, the probability density of

Δm makes it possible to obtain σ. Parameters α and β are deduced from the correlation law between a and m given in [3.17], which is calculated from the same experimental data.

So, from this data, we need to deduce $\tilde{\mu}_m$ and $\tilde{\sigma}_m$. Note that the value of $\tilde{\mu}_m$ can also be calculated directly by averaging m.

Recall that we are now trying to calculate the distribution law of failure time (t_{eol}). For this, we set its definition through equation [3.30] and the joint probability density of a and m, $f_{A,M}(a,m)$, given in [3.22] and [3.25] is recalled in equation [3.31].

$$t_{eol} = \sqrt[m]{\frac{\delta}{a}} \qquad\qquad [3.30]$$

$$f_{A,M}(a,m) = g_A(a) \times n_M(m - \alpha \ln(a) - \beta) = a^{\alpha} \times g_A(a) \times g_{A^{\alpha}}(a^{\alpha}) \qquad [3.31]$$

The proposed approach here is identical to the one used for the calculation of $f_{A,M}(a,m)$, corresponding to application of a change of variable.

We then write the new coordinate system as follows:

$$\binom{a}{m} \mapsto \begin{cases} t_{eol-ln} = \dfrac{1}{m}\ln\left(\dfrac{k}{a}\right) \\ m = m \end{cases} \Leftrightarrow \begin{cases} a = k\ e^{-mt_{eol-ln}} \\ m = m \end{cases} \qquad [3.32]$$

We calculate the reverse Jacobian operator as follows:

$$J^{-1} = \begin{vmatrix} -mke^{-mt_{eol-ln}} & -t_{eol-ln}ke^{-mt_{eol-ln}} \\ 0 & 1 \end{vmatrix} = -mke^{-mt_{eol-ln}} \qquad [3.33]$$

We can then write the failure times distribution in the following form:

$$f_{T_{eol},M}(t_{eol},m) = |m|\,ke^{-mt_{eol-ln}}g_{T_{eol}}(ke^{-mt_{eol-ln}})n_M(m) \qquad [3.34]$$

with:

$$\begin{cases} g_{T_{eol}}(ke^{-mt_{eol-ln}}) \\ n_M(m) \end{cases} \text{ having as parameters respectively: } \begin{cases} l_a,s \\ -\alpha\ln(ke^{-mt_{eol-ln}}) + \beta, \sigma \end{cases} \quad [3.35]$$

Thus, we can have the distribution function of t_{eol}, $F(t_{eol})$ by integrating equation [3.34] with respect to the variable m over the whole of its domain and with respect to t_{eol} over the interval [0, t], in which t is the chosen end-of-life time.

3.4.1.2. Application of theoretical calculation

This analytical calculation was applied to the eight experimental couples (a, m) already used in section 3.3.3; the objective of this section is to demonstrate the validity of the Monte Carlo method used in our reliability estimation methodology.

Having previously developed the theoretical approach, we will only present here the main results related to this application without going in details. First, we will recall the parameters necessary to be used in equation [3.17], as well as those defining the distributions of parameters a and **Δm** in Table 3.5.

Studied law	Parameters of each law
Correlation law (a,m)	$\alpha = -0.0647$ $\beta = 0.2422$
Distribution of "a"	$l_a = 0.061$ $s = 1.679$
Distribution of "Δm"	$m_0 = 0$ $\sigma = 0.045$

Table 3.5. *Definition of the useful parameters for the analysis*

From these parameters and according to the steps described previously, we define the joint probability density of the random variables T_{eol} and M, as $f_{T_{eol},M}(t_{eol},m)$. This function is shown in Figure 3.9. Then, the probability density of the failure times (t_{eol}) and their distribution functions are, respectively, shown in Figures 3.10 and 3.11.

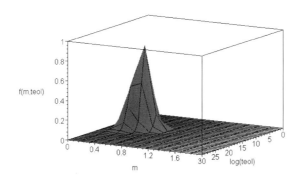

Figure 3.9. *Probability density of $f_{teol,m}(t_{eol},m)$. For a color version of this figure, see www.iste.co.uk/vanzi/reliability.zip*

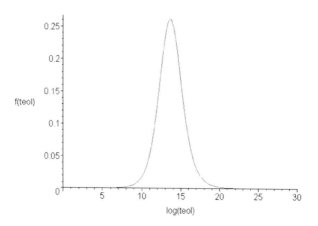

Figure 3.10. *Probability density of t_{eol}*

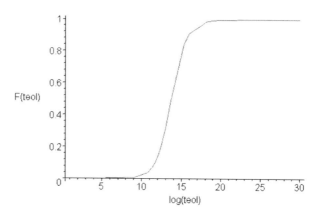

Figure 3.11. *Distribution plot of $[t_{eol}, F_{teol}(t_{eol})]$ (on the abscissa, logarithmic scale)*

Finally, the failure rate can be extrapolated at 20 years, λ_{20}, by considering the following equation:

$$\lambda_{20} = \frac{F_{T_{eol}}(t = 20\,\text{years})}{t(1 - F(t))} \times 10^9$$

[3.36]

Calculation method	Monte Carlo	Analytical
λ_{20} in FITs (accelerated conditions, AF = 8)	930	990
λ_{20} in FITs (operating conditions)	116	124

Table 3.6. *Failure rate at 20 years (in FIT) in accelerated conditions (AF = 8) and in operating conditions: comparison between the analytical method and the Monte Carlo method (1024 samples)*

The value λ_{20} thus obtained is compared with the one calculated from the Monte Carlo method. The difference between the two methods is around 5% at most, having in mind that the analytical method requires a much greater operator involvement, because the calculation of integrals to be carried out are not trivial and require simplifications and other variable changes, not to mention the optimization of the areas of integration. As part of more systematic or even automated studies (implementation as part of a qualification process), the valuable use of the Monte Carlo process in the methodology is strongly recommended.

3.4.2. Robustness analysis of the statistical random draws

3.4.2.1. Effect of a bias in the correlation law

The complete description of the methodology can appear, in a very simplified form, as represented in Figure 3.12. However, this figure shows that the experimental couples (a, m) represent the only inputs. Also, none of the basic steps can minimize an error that would be made in the extraction of these couples. Therefore, an error on the first element of the chain necessarily results in a cascade of errors on the other elements. The purpose of this section is therefore focused on the evaluation of the impact of the error made on this first element of the chain by the user. We will endeavor to analyze the adjustment of the experimental curves by the laws in at^m, on the latter, which correspond to the calculation of $\lambda(t)$ for a time equivalent to the estimated (or desired) lifetime of the device.

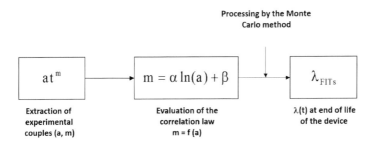

Figure 3.12. *Simplified processing diagram*

First, we will:

1) assess the impact of the bias on the correlation law in relation to the result of the simulation;

2) quantify the error on the extraction of experimental couples (a, m) corresponding to this same bias;

3) make a parallel between this error and the result in FITs obtained at the end of the processing chain.

For the first step, namely taking into account of a bias on the correlation law, m = f (a), the modification of the function f is carried out through a modification of the experimental couples (a, m). The following three types of modification are successively carried out:

– a direct modification of the value of m by adding a bias represented by a constant (δ), whatever the value m. This induced error, which is generally unlikely, has the advantage of only varying the ordinate at the origin of the correlation, leading to a de-correlation between the modifications of α and those of β;

– again, a direct modification of the value of m, but the bias P_m corresponds to a percentage of m. This modification of m then induces both a change in the directing coefficient α of the correlation and of its ordinate at the origin β;

– finally, a modification of the error Δm, committed on the calculation of the value of m when it is calculated from the correlation law and not obtained experimentally. This modification results in the addition of a value Δm, corresponding either to a percentage of the value of Δm, or to a constant whatever the value of Δm.

First case: introduction of a constant error in the correlation law

The first technique consists of inducing a bias on the correlation law by adding a quantity δ to this law, as shown in equation [3.37]. However, this quantity δ being constant, equation [3.37] can then be replaced by equation [3.38], where this added quantity is integrated into the ordinate at the origin β, expressing here the fact that this bias will arise in the form of a translation of the correlation curve along the axis of m (Figure 3.13).

$$m = \alpha \ln(a) + \beta + \Delta m + \delta \qquad [3.37]$$

$$m = \alpha \ln(a) + \beta' + \Delta m \qquad [3.38]$$

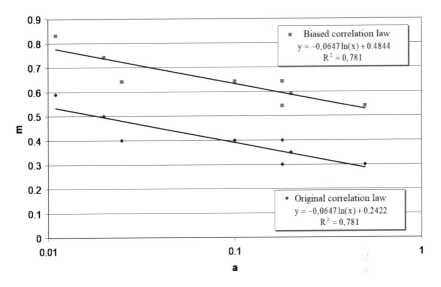

Figure 3.13. *Initial correlation law and biased one after introduction of a constant error. For a color version of this figure, see www.iste.co.uk/vanzi/reliability.zip*

The results of such a modification on the simulation are summarized in Table 3.7, thus making it possible to compare the failure rates at 20 years and the distribution laws of the variables **a** and **Δm**, for the case where $\delta = 0$ and $\delta = \beta$. It was expected that no change would appear in these distribution laws. Indeed, we see with equation [3.38] that the variables undergoing random sampling are not impacted by the modification in the correlation law, and therefore, the processing chain should, *a priori*, undergo no consequence.

Quantitative bias	Parameters considered in the distribution laws of a and Δm	
	"a" law	"Δm" law
$\delta = 0$	m = 0.0607 s = 1.679	$\mu = 0$ $\sigma = 0.0455$
$\delta = \beta$	m = 0.0607 s = 1.679	$\mu = 0$ $\sigma = 0.0455$

Table 3.7. *Comparison of simulation results without bias ($\delta = 0$) and with bias ($\delta = \beta$)*

However, we note that variations appear when we construct the joint distribution $F_{a,m}$ and the calculation of the failure times. This is justified by the fact that it is necessary to reconstruct the distribution of **m** from the couples **(a,Δm)** created by the drawing process, by reintroducing them into the formula [3.38]. We can thus observe this difference between the first failure times between the two simulations in Figure 3.14 representing the rate of cumulative failures as a function of time.

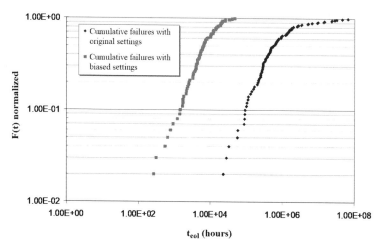

Quantitative bias	$\delta = 0$	$\delta = \beta$
λ_{20} in FITs (accelerated conditions, AF = 8)	980	14900
λ_{20} in FITs (operating conditions)	122	1860

Figure 3.14. *Cumulated failure rates versus time calculated without bias ($\delta = 0$) and with bias ($\delta = \beta$) error. For a color version of this figure, see www.iste.co.uk/vanzi/reliability.zip*

The effect of an increase in the ordinate at the origin β, in the correlation law involves an earlier appearance of the first instants of failure (two decades compared to t_{eol} of the law of original correlation).

This first study therefore makes it possible to quantify the impact of a simple error on the smoothing of the correlation law. However, as the error presented here seems unnatural, other types of bias had to be studied, which could be interpreted as reading errors or taking into account an error made in the evaluation of the couple (a, m) of one of the devices of the studied whole batch, thereby distorting the entire test batch.

Therefore, we have introduced two additional types of bias:

– a first one depending on the parameter m;

– and a second one depending on the error between the correlation and each experimental point.

Second case: introduction of an m-dependent error in the correlation law

This second study is based on an approach quite similar to the previous one, namely the addition of a quantity δ to the value of the experimental m. The difference here is in the nature of δ. In the previous study, we considered δ as a constant regardless of the value of m; in this case, **the bias will be represented as a percentage P of the value of m to which it is added** (equation [3.39]). We then obtain:

$$m = \alpha \ln(a) + \beta + \Delta m + P_m$$
$$\text{with } P_m = P \times m$$

[3.39]

We consider that the percentage P is constant whatever the value of m. By adding the quantity P_m to the correlation equation, the original correlation law is then weighted by the quantity $1/P_m$. Thus, new values, denoted as α' and β', will appear in equation [3.40], but also a value $\Delta m'$, thus modifying the parameters of the distributions calculated in the processing chain, as presented in Figure 3.15.

$$m = \alpha' \ln(a) + \beta' + \Delta m'$$

[3.40]

The simulation results related to this study are given in Table 3.8 for a P_m value of 100%. These results are also compared with the simulation results obtained for the original pairs **(a, m)**, considered as unbiased.

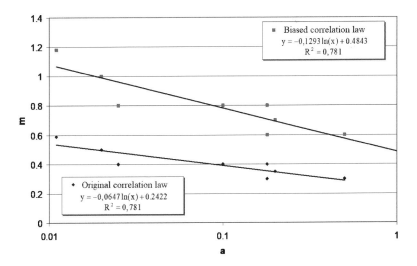

Figure 3.15. *Initial correlation law and biased one after introduction of an m-dependent error. For a color version of this figure, see www.iste.co.uk/vanzi/reliability.zip*

Quantitative bias	Parameters considered in the distribution laws of a and Δm	
	"a" law	"Δm" law
P = 0	m = 0,0607 s = 1,679	μ = 0 σ = 0,0455
P = 100 %	m = 0,0607 s = 1,679	μ = 0 σ = 0,0910

Table 3.8. *Comparison of simulation results without bias*
(P = 0, δm = 0) and with bias (P = 100%, δm = m) error

Figure 3.16 highlights the distribution of Dm (*density of probability of presence of Dm in the interval [-0.5, +0.5]*) for sampling processes of original couples (a, m) and biased ones. The distribution of the cumulative failure rates considering original and biased couples is plotted in Figure 3.17, highlighting a difference between the appearance of the first failure times of more than two decades, that is a more pessimistic result than those highlighted with the previous study. The failure rates a 20 years are given in the table below Figure 3.17, showing a significant increase (a decade and a half). The difference between the failure rate calculated from the original couples (a, m) and from the biased couples is 5 times greater than that of the previous case, where only the value of b was modified.

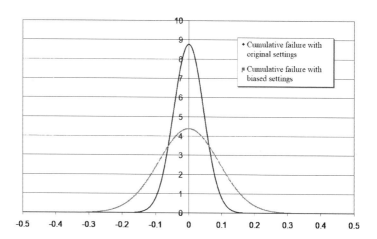

Figure 3.16. *Distribution of Δm for sampling processes of original (a, m) numbers and biased ones. For a color version of this figure, see www.iste.co.uk/vanzi/reliability.zip*

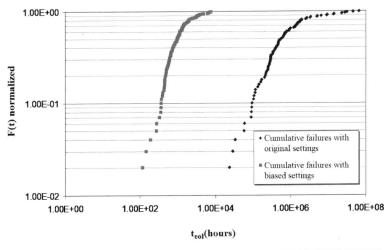

Quantitative bias	P = 0	P = 100%
λ_{20} in FITs (accelerated conditions, AF = 8)	980	56,200
λ_{20} in FITs (operating conditions)	122	7,025

Figure 3.17. *Cumulated failure rates versus time calculated without bias (P = 0, δm = 0) and with bias (P = 100%, δm = m) error. For a color version of this figure, see www.iste.co.uk/vanzi/reliability.zip*

This increase in the differences between the simulations performed from biased and non-biased couples can be explained by the modification of additional parameters in the correlation equation. However, in this specific case, two parameters were modified at the same time. An additional study, carried out hereafter, is therefore necessary to discriminate the most critical parameter responsible for this difference.

Third case: introduction of a Dm-dependent error in the correlation law

In this part, the introduced bias is directly related to the error on the correlation law between a and m, and this through a modification of Δm. Equation [3.41] illustrates this modification, thus enabling us to isolate the variation of this parameter compared to the previous study. In the case that we will present here, and by way of example, the error Δm has been multiplied by 2. As shown in Figure 3.18, the correlation law thus obtained differs from the initial law only by the value of the linear regression coefficient R^2.

$$m = \alpha \ln(a) + \beta + \Delta m' \qquad [3.41]$$

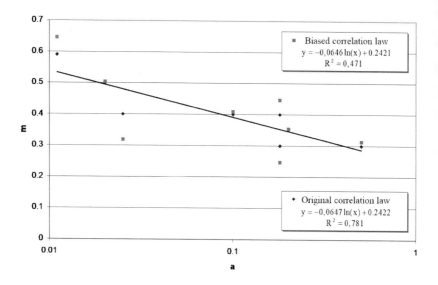

Figure 3.18. *Initial correlation law and biased one after introduction of a Δm-dependent error $\left[\Delta m' = 2 \times \Delta m\right]$. For a color version of this figure, see www.iste.co.uk/vanzi/reliability.zip*

Table 3.9 presents the simulation results in terms of failure rate as well as the distribution parameters of the two variables a and **Δm**. Note that the failure rates, with and without error bias, are of the same order of magnitude despite the increase in the standard deviation of the correlation error, which can also be seen in Figure 3.19 representing the density of probability of presence of **Δm** in the interval [−0.5, +0.5].

Quantitative bias	Parameters considered in the distribution laws of a and Δm	
	"a" law	"Δm" law
$\Delta m' = \Delta m$	m = 0,0607 s = 1,679	μ = 0 σ = 0,0455
$\Delta m' = 2 \times \Delta m$	m = 0,0607 s = 1,679	μ = 0 σ = 0,0911

Table 3.9. *Comparison of simulation results for Δm′ = Δm and Δm′= 2 × Δm*

Figure 3.19 plots the distribution of Dm (*density of probability of presence of Dm in the interval [-0.5, +0.5]*) for sampling processes of original couples (a, m) and biased ones. With Figure 3.20, one can note that the failure rates, without and with bias, are quite often of the same order of magnitude.

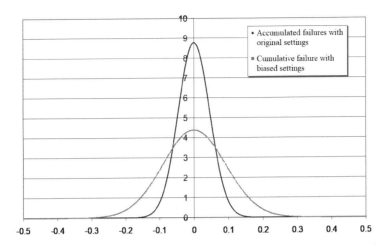

Figure 3.19. *Distribution of Δm for sampling processes of original (a, m) numbers and biased ones. For a color version of this figure, see www.iste.co.uk/vanzi/reliability.zip*

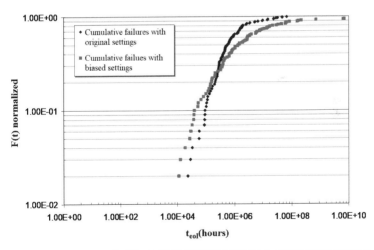

Quantitative bias	$\Delta m' = \Delta m$	$\Delta m' = 2 \times \Delta m$
λ_{20} in FITs (accelerated conditions, AF = 8)	980	920
λ_{20} in FITs (operating conditions)	(122)	(115)

Figure 3.20. *Cumulated failure rates versus time calculated without bias (Δm' = Δm) and with bias (Δm' = 2 ×Δm) error. For a color version of this figure, see www.iste.co.uk/vanzi/reliability.zip*

Identically to the failure rates, Figure 3.20 gives a comparison between the cumulative failure rates as a function of time obtained from the original and biased settings, showing that the first failure times are of the same order of magnitude.

From this study, we note that a bias, even significant, on the error Δm of the correlation law has little influence on the simulation results. This last result confirm that the use of the parameter Δm as an independent variable of a, to achieve a double Monte Carlo draw, does not introduce any additional error in our approach.

3.4.2.2. Summary

Through this study, we have determined the impact of a bias introduced into the correlation law on the cumulative failures rate. The purpose of this approach was to verify the influence of the error made when determining the correlation law between the two parameters a and m. We were able to demonstrate that:

– **A bias on the magnitude of the correlation error (Δm) has a negligible influence on the simulation results.** This result confirms our approach. We can therefore replace the variable m, a non-random variable because it is correlated with the variable a. Now we cannot apply, to this variable, the Monte Carlo method in order to achieve a joint distribution, by a random variable Δm resulting from the dissociation of the part of m fully correlated to a, called $\tilde{m}$, and from the error made with respect to the correlation law.

– **An error on the parameters α and β of the correlation law during its establishment has a significant influence on the calculation of the failure rate $\lambda(t)$ and on the first moments of occurrence of failures.** An optimized and systematic approach to determining the correlation law itself, as well as a reliable approach for extracting the couples (a, m) that are at their origin, therefore remain essential to implement.

This robustness study is obviously not exhaustive. It only takes into account the biases, which we assume to be the most important, of a given correlation law, but could be widened in particular to the choice of another type of correlation law (exponential or other). However, it allows us to assess the criticality of the biases introduced in the initial step, a fundamental step in our processing chain.

3.4.2.3. Correlation law optimization

As we have just shown, the adjustment on the correlation law is a crucial step of this methodology. Also, in order to ensure maximum precision, we will evaluate the relevance of different estimators compared to the power law used to deduce the evolution of parameters on weak test drifts. Three estimators were studied:

– empirical estimator;

– least squares estimator;

– maximum likelihood estimator.

The main hypothesis is that the degradation kinetics can be written in the following general form:

$$d = at^m$$

d : decay amplitude [3.42]

t : time

1) Empirical estimator: $\ln d = \ln a + m \ln t$

We do here a linear regression, on N points, of $Y = \ln d$ according to $X = \ln t$. The parameter noted $\tilde{m}$ is then obtained from the estimated value of m, considering the following approach:

$$\tilde{m} = \frac{\text{cov}_{XY}}{\text{var}_X} = \frac{\displaystyle\sum_{1}^{N}\left(\ln d_i - \overline{Y}\right)\left(\ln t_i - \overline{X}\right)}{\displaystyle\sum\left(\ln t_i - \overline{X}\right)^2}$$

[3.43]

with: $\overline{Y} = \dfrac{1}{N}\displaystyle\sum_{1}^{N}\ln d_i$ and $\overline{X} = \dfrac{1}{N}\displaystyle\sum_{1}^{N}\ln t_i$

Likewise, the parameter $\tilde{a}$ is then obtained from equation [3.44]:

$$\tilde{a} = e^{\overline{Y} - \tilde{m}\overline{X}}$$

[3.44]

The advantages of this estimator lie in its simplicity and ease of use. The disadvantages reside in the fact that it is biased by the change of variable and that it can be faulted for negative values of d, emphasizing a great sensitivity to noise.

2) Least squares estimator

This estimator allows us to minimize the variance expressed by:

$$e = \sum\left(d_i - at_i^m\right)^2$$

[3.45]

We then estimate the values $\tilde{a}$ and $\tilde{m}$ of variables a and m using equations [3.46] and [3.47]:

$$\tilde{a} = \frac{\sum_1^N d_i t_i^{\tilde{m}}}{\sum_1^N t_i^{2\tilde{m}}} \tag{3.46}$$

$$\tilde{m} \left| \sum_1^N t_i^{2m} \sum d_i t_i^m \ln t_i - \sum d_i t_i^m \sum t_i^{2m} \ln t_i = 0 \right. \tag{3.47}$$

This method is very robust with regard to negative values. However, the implicit equation for m makes the method slow for N large and remains sensitive to the addition of strong noise.

Maximum likelihood estimator

This estimator has the property of being insensitive to changes in the variable, and of having, as an estimator of the mathematical expectation of x (denoted E {x}), the mean value of x (denoted $\overline{x}$).

In this case, when we set $X = \dfrac{d}{a t_i^m}$, we must obtain E {X} = 1, that is:

$$\frac{1}{N} \sum_1^N \frac{d_i}{t_i^{\tilde{m}}} = \tilde{a} \tag{3.48}$$

Similarly, by letting $Y = \ln d - \ln a - m \ln t$, this results in E {Y} = 0, and equation [3.49] corresponds to a likelihood equation making it possible to deduce m knowing a to be maximized by solving [3.50].

$$\sum_1^N (\ln d_i - \tilde{m} \ln t_i) - N \ln \tilde{a} = \overline{\ln d_i} - \tilde{m} \overline{\ln t_i} - N \ln \left(\frac{1}{N} \sum_1^N \frac{d_i}{t_i^{\tilde{m}}} \right) \tag{3.49}$$

$$\overline{\ln t_i} \sum_1^N \frac{d_i}{t_i^{\tilde{m}}} - \sum \frac{d_i \ln t_i}{t_i^{\tilde{m}}} = 0 \tag{3.50}$$

This method has the advantage of being very robust against negative values. However, several disadvantages appear – it is an implicit equation for m and the reasoning must be constrained by strong assumptions on the law of evolution. The considered estimator cannot therefore appear to be a true estimator in the sense of maximum likelihood.

Given the low number of points N describing the experimental evolution of the parameters studied on our components and the simplicity of use of the method, **our choice therefore focused on the least squares estimator.**

3.4.2.4. Failure rate distribution

The goal here is to analytically reconstruct the distribution of failure rates over time. The calculation made from the t_{eol} obtained thanks to the random drawing process, and the cumulative failure function (see equation [3.37]) is adjusted using a distribution function. The choice of the nature of this function (log-normal or Weibull) must naturally be user dependent. However, the difference in behavior between these two functions requires studying the impact of this choice on the result when calculating failure rate at a given time; in our case, at 20 years.

A comparative test was therefore carried out using the couples used during the demonstration of the method by the example proposed in section 3.3.3. On several occasions, the failure rate at 20 years (λ_{20}) has been calculated by the two different functions, with the virtual couples (a, m) of a one and same drawing each time. The results are summarized in Table 3.10 through an example, as well as the parameters associated with the chosen distribution laws. Figure 3.21 presents the comparison of the default rate adjustment by considering two types of law (refer to Table 3.4 for a definition):

– the log-normal law;

– Weibull's law.

Distribution	Failure rate in accelerated conditions (and in operating conditions) at 20 years (in FITs)	Parameters considered in the distribution laws
log-normal	965 (120)	m = 866500 s = 1,39
Weibull	725 (90)	$\alpha = 1307350$ $\beta = 1,053$

Table 3.10. *Prediction of failure rate, λ(t), using different distributions (log-normal and Weibull)*

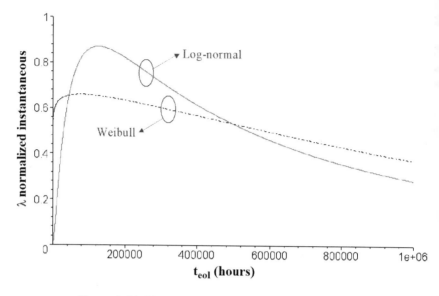

Figure 3.21. *Mathematical fitting of the failure rate, $\lambda(t)$, using log-normal and Weibull distributions*

Regarding the result of calculating the failure rate at 20 years, the choice of Weibull's law compared to a log-normal law does not seem to present a significant difference, since the results remain of the same order of magnitude. However, it should be noted that the adjustment by Weibull's law systematically underestimates the λ_{20} compared to the one by a log-normal law.

For the sake of consistency, the user must choose an adjustment law and keep it throughout the processing chain. **Our choice has been focused on the log-normal law for two main reasons**:

– the maximum of the log-normal fitting curve is delayed compared to that of Weibull fitting. Therefore, the first failure times are not originally rejected, as in the case of Weibull's law, which would not be very compatible with physical reality in our case;

– the increase in the result, compared to Weibull's law, gives a failure rate certainly more pessimistic at 20 years, but ensuring a usable upper in the qualification process.

3.5. Experimental investigations

3.5.1. *Single parameter estimation: drift of the bias current during aging tests*

Parameter I_{Bias} is one of the most frequently used parameters, associated with the variation of optical power as a function of the current, in terms of characterization of the behavior of a laser diode during reliability investigations (Hwang *et al.* 1997; Resneau 2004). This parameter is a first-order parameter since the direct bias of a laser is done by current (perfectly stabilized), but also to the aging test technique used. Moreover, during an aging test conducted at constant emitted power (known as "APC" while "ACC" refers to a test at constant current), the current can be easily monitored over the test duration (Goudard 2000).

Based on this observation, we applied our methodology to a specific batch of 20 DFB 1550 nm laser diodes, from the 30 components used previously, with the aim of reconstructing the distribution of failure times and performing a prediction of failure rate at 15 years (for a terrestrial application) by applying the statistical approach proposed above, in particular by taking into account two major limitations:

– the small population of components;

– small drifts that necessarily require extrapolation.

Indeed, after 10,000 h of aging at 53°C and 110 mA (AF = 5), the results are as follows:

– 14 laser diodes have variations in supply current of less than 2%;

– six laser diodes show variations in I_{Bias} between 2% and 10%.

These variations are therefore far below the 20% failure criterion recommended by the TELCORDIA GR-468 standards for this parameter (Chuang *et al.* 1997; Telcordia GR-468-CORE 1998).

From the aging results thus obtained, the variations are modeled by at^m type fitting curves. The determination of the couples (a, m) thus carried out then makes it possible to apply the methodology by Monte Carlo drawing to obtain a statistical reinforcement from 1024 virtual couples (a, m) in order to reduce the global error on the failure rate thus extrapolated in operating conditions. The cumulative failure rate as a function of time is presented in Figure 3.22, making it possible to determine the first instant of failure around 130,000 operating hours, which represents 14.8 years of operation. The 15-year failure rate is estimated at 300 FITs.

This failure rate is lower than that recommended for use in a terrestrial optical telecommunication network, namely 500 FITs (Sauvage *et al.* 2000).

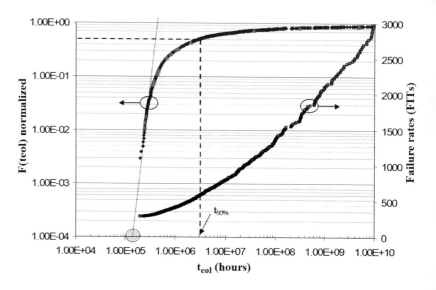

Figure 3.22. *Failure rate distribution versus time calculated from 1024 virtual couples (a, m) after Monte Carlo random processing (in blue) including 20 experimental couples (a, m) (in red) (Béchou 2015). For a color version of this figure, see www. iste.co.uk/vanzi/reliability.zip*

3.5.2. *Multiple parameters estimation: drifts of threshold current and optical efficiency*

The study of the I_{Bias} parameter presents a major disadvantage. The main hypothesis lies with the fact that the degradation mechanisms observed, described by their kinetics, will be the only ones that will appear during non-premature aging at the end of the test. Due to the short test times and the low number of components, this hypothesis must be reconsidered through a variation of several parameters on which the bias current I_{Bias} depends (namely, the threshold current I_{th} and the optical efficiency α).

Note that these parameters are linked by equation [3.51]. The change in optical power can be written by equation [3.52]. Operating at constant power makes it possible to simplify the calculation by writing equation [3.53].

$$P_{opt} = \alpha \left(I_{Bias} - I_{th} \right) \qquad [3.51]$$

$$dP_{opt} = \left(I_{Bias} - I_{th}\right)d\alpha - \alpha dI_{th} + \alpha dI_{Bias} = 0 \qquad [3.52]$$

$$dI_{Bias} = -\frac{\left(I_{Bias} - I_{th}\right)}{\alpha}d\alpha + dI_{th} \qquad [3.53]$$

Relative variations, more suited to experimental data, are calculated in [3.54], and can be simplified as reported in [3.55]. The coefficients K_1 and K_2 are directly determined by identification between equations [3.54] and [3.55]. It is then noted that these coefficients are such that $K_1 + K_2 = 1$ with $K_1 > K_2$ (if the component is biased sufficiently above the threshold) (Mendizabal 2004):

$$\frac{\Delta I_{Bias}}{I_{Bias0}} = -\left(\frac{I_{Bias0} - I_{th0}}{I_{Bias0}}\right)\frac{\Delta\alpha}{\alpha} + \left(\frac{I_{th0}}{I_{Bias0}}\right)\frac{\Delta I_{th}}{I_{th0}} \qquad [3.54]$$

$$\frac{\Delta I_{Bias}}{I_{Bias0}} = -K_1\frac{\Delta\alpha}{\alpha} + K_2\frac{\Delta I_{th}}{I_{th0}} \qquad [3.55]$$

Extrapolating the variations in I_{Bias} obtained from the drift in accelerated aging of the devices used in this study allows the coefficients K_1 and K_2 to be evaluated at 0.75 and 0.25, respectively, verifying the condition $K_1 > K_2$. The influence of the variations of α and I_{th} through the variations of I_{Bias} was studied on the six most degraded devices in order to distinguish the behavior of each parameter. Two modes of degradation have been pointed out:

– **mode no. 1**: this first mode of degradation, appearing on three lasers, presents a rapid increase in I_{Bias} associated with an increase in I_{th} together with a drop in α;

– **mode no. 2**: the second mode, observed on the three remaining components, presents a gradual increase in I_{Bias} only associated with a significant increase in I_{th}.

The variations of α and I_{th} for cases 1 and 2 are presented, respectively, in Figures 3.23 and 3.24. In general, the analysis of the physical mechanisms associated with these degradation modes (see previous chapters) made it possible to link these two cases to two types of degradation which are widely detailed and discussed in the literature on the topic, see for example (Fukuda 1991):

– **mode no. 1** corresponds to an increase in the rate of non-radiative recombinations in the active area, associated with an increase in internal losses within the cavity (increase in the absorption of photons);

– **mode no. 2** can be associated with a single increase in the rate of non-radiative recombinations.

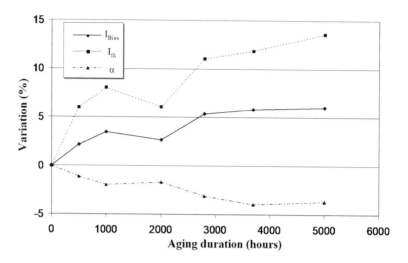

Figure 3.23. Drift of I_{Bias}, I_{th} and α corresponding to failure mode no. 1

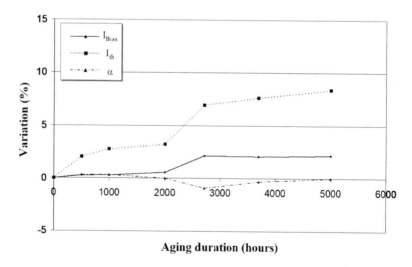

Figure 3.24. Drift of I_{Bias}, I_{th} and α corresponding to failure mode no. 2

For each case, the cumulative failure functions were constructed from statistical data on I_{Bias} using the proposed methodology based on Monte Carlo draws (see Figure 3.25). An extrapolation is then carried out using the log-normal type draw to determine the first failure times.

In Figure 3.25, these extrapolations are shown in solid lines for case no. 1 and in dotted lines for case no. 2. Thus, in case no. 1, the first failure times are extrapolated around 3×10^4 hours compared to 1.5×10^5 hours for the second case. This difference was foreseeable given the significant differences between the variations in I_{Bias} for the two cases.

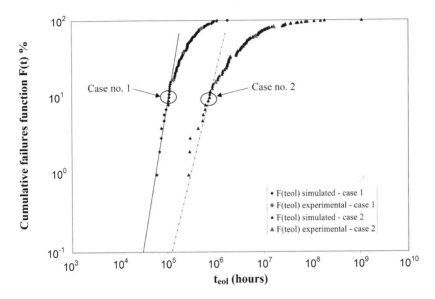

Figure 3.25. *Prediction of failure rate distribution versus time calculated from 1024 virtual couples (a, m) after Monte Carlo random processing (in blue) including 20 experimental couples (a, m) (in red) related to the drift of I_{Bias} (single parameter estimation). For a color version of this figure, see www.iste.co.uk/vanzi/reliability.zip*

The multiparametric study of I_{th} and α variations shows that it can provide additional information compared to the monoparametric study based solely on I_{Bias} due to the difference between the degradation mechanisms (modes nos 1 and 2).

The Monte Carlo method was therefore applied to the I_{th} parameter, taking a variation of 20% as failure criterion required by TELCORDIA standards (Telcordia 1998). The results of these simulations are presented in Figure 3.26 in the form of two cumulative failure functions for the two cases.

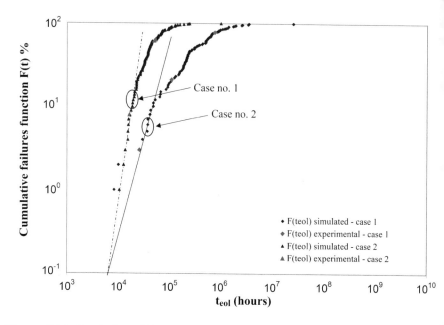

Figure 3.26. *Prediction of failure rate distribution versus time calculated from 1024 virtual couples (a, m) after Monte Carlo random processing (in blue) including 20 experimental couples (a, m) (in red) related to the drift of I_{th} (single parameter estimation). For a color version of this figure, see www.iste.co.uk/vanzi/reliability.zip*

Figure 3.26 clearly shows that the first failure times obtained appear earlier (around 8×10^3 h) than those obtained in case no. 1 from the cumulative failure function of I_{Bias}.

The difference between the two simulations is mainly related to the experimental distribution of the parameters before aging and to the weight of the constant K_2, as expressed in equation [3.55]. The impact of the variations of α is quantifiable by considering the difference of the first failure times of cases no. 1 and 2, calculated for the parameter I_{Bias} (Figure 3.25).

In fact, the behavior observed in case no. 2 is considered to be a precursor of the decay described in case no. 1. Indeed, the bibliography has clearly established that a fairly significant increase in non-radiative recombinations in the active area can lead to damage to optical efficiency through a significant absorption of photons (Park *et al.* 2003). Particular attention must therefore be paid to the fact that the extrapolation of the data of mode no. 2, obtained by simulation of the I_{Bias} parameter, can lead to optimistic results considering simply the shifts of the t_{eol}, and the possibility of an increase in the parameter a in equation [3.1], defining the calculation of the failure times according to the couples (a, m) as well as the failure criterion.

This increase, due to a sudden change in optical efficiency, remains very difficult to demonstrate during short-term aging tests. It is therefore necessary, when analyzing the I_{Bias} parameter, to study the I_{th} and α parameters in a complementary manner. Indeed, we have shown that the first failure times obtained by the study of I_{Bias} were overestimated compared to those resulting from the study of I_{th}. Thus, the observation of the behavior of I_{Bias} alone does not account for the progress of a degradation of I_{th}. Figure 3.27 presents the comparison of the cumulative failure functions of I_{Bias} and I_{th} only in the context of case no. 2; failures are not related to α. **However, the optimistic nature of the variations in I_{Bias} must be taken into account for some "I_{th} – dependent" applications (i.e. operation in direct modulation, for example).**

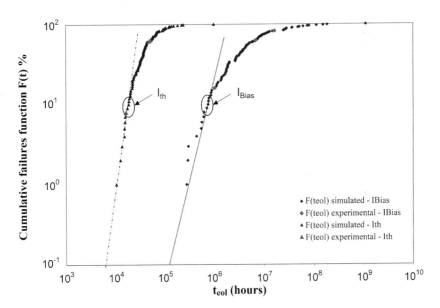

Figure 3.27. *Prediction of failure rate distribution versus time calculated from 1024 virtual couples (a, m) after Monte Carlo random processing (in blue) including 20 experimental couples (a, m) (in red) related to the drift of I_{Bias} and I_{th} (multiple parameters estimation – mode no. 2). For a color version of this figure, see www.iste. co.uk/vanzi/reliability.zip*

Finally, the relevance of this reasoning is reinforced through the application of equation [3.55] to calculate t_{eol} obtained during studies of the different parameters I_{Bias}, I_{th} and α. Figures 3.28 and 3.29 present the result of this verification by comparing the distribution of the failure times applied to each member of equation [3.55] and which are based on the variations of the parameters I_{Bias}, I_{th} and α reported in Figures 3.23 and 3.24 for modes nos 1 and 2, respectively. This approach is achievable by considering that the m parameters of each variation are very close. The observable differences between the cumulative failure functions of I_{Bias} and those obtained by linear combination of the functions related to α and I_{th} are mainly due to the fact that each $F(t_{eol})$ of each parameter comes from a different random drawing, generating its own random error.

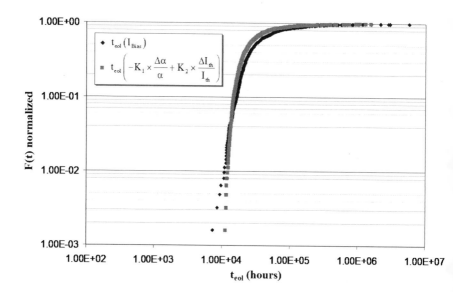

Figure 3.28. Prediction of failure rate distribution versus time calculated from 1024 virtual couples (a, m) after Monte Carlo random processing (in blue) including 20 experimental couples (a, m) (in red) related to the drift of I_{Bias} only and considering I_{Bias} as a linear combination of (α, I_{th}) (multiple parameters estimation – mode no. 1 – K_1 and $K_2 > 0$). For a color version of this figure, see www.iste.co.uk/vanzi/reliability zip

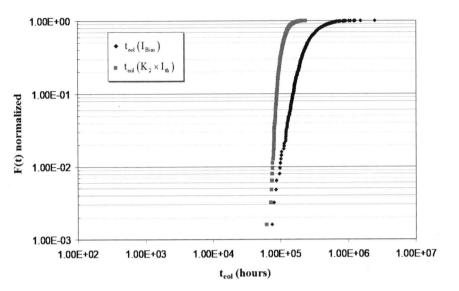

Figure 3.29. *Prediction of failure rate distribution versus time calculated from 1024 virtual couples (a, m) after Monte Carlo random processing (in blue) including 20 experimental couples (a, m) (in red) related to the drift of I_{Bias} only and considering I_{Bias} as a linear combination of (α, I_{th}) (multiple parameters estimation – mode no. 2 – $K_1 \sim 0$ and $K_2 > 0$). For a color version of this figure, see www.iste.co.uk/vanzi/reliability.zip*

3.5.3. *Summary*

We have demonstrated the relevance of the analysis of additional parameters to I_{Bias} as part of a study of the reliability of 1550 nm DFB laser diodes. Indeed, the study of different parameters brings us greater assurance in the extrapolation than just taking into account the I_{Bias} parameter, and makes it possible to assess the reliability of the device in the context of specific applications. In particular, we used the example of the I_{th} parameter, where a too early drift could be problematic in an application requiring direct modulation of the laser diode. However, we could have also considered the drift of the emission wavelength or spectral properties of the laser diode for wavelength-division multiplexing (WDM) applications as proposed by (Sobiestiankas 2005).

The reliability study from a multiparametric approach makes it possible to propose a close link with the physical mechanisms of degradation. We were thus able to distinguish two failure modes, denoted as mode nos 1 and 2, within the batch of laser diodes emitting at 1550 nm, characterized by:

– **mode no. 1: increase in the laser threshold current I_{th}**, in relation to an increase in the rate of non-radiative recombinations, **and decrease in the optical efficiency α**, caused by an increase in internal losses;

– **mode no. 2: increase in the threshold current I_{th} only.**

Let us recall that these behaviors, demonstrated experimentally and exhaustively reported in the literature on similar technologies, shows that case no. 1 is identified as being the consequence, of case no. 2 in the long-term.

Generally, a batch of components under test very rarely presents a unique failure signature. We often highlight a certain percentage of components whose behavior corresponds to case no. 1 and another representative percentage of case no. 2. This batch will therefore show an evolution toward uniformity, more or less long term, relating to the degradations of case no. 1. Lifetime predictions of this batch, through the monoparametric analysis, will therefore lead irreparably to an optimistic conclusion, as indicated in Figure 3.29. The study, carried out by considering case no. 1-like degradations, based on a multiparameter approach, therefore gives a confidence interval, giving a maximum limit in terms of lifetime of this batch.

It should also be noted that, in order to discriminate failure signatures, one could also carry out the opposite approach and show from what point the degradations related to α are likely to appear. Thus, a statistical study of the degradation part only related to α, in a batch of devices, would make it possible to ensure greater relevance to the study of the failure times.

However, a delicate point remains: to take into account of the degradation mechanisms in a reliability study, when the law used to fit the variations of the parameters studied, corresponds to an empirical law, that is not always representative of physical behavior. In order to investigate more about the physics of failures, it will therefore be necessary to use an adjustment law that is more connected to the technology of the device.

As such, the last part of this chapter aims to briefly describe different physical models that can be used to extrapolate the variations of the I_{th}, allowing us to link with the next chapter.

3.6. Toward multi-components physical models

As we specified in section 3.1, the so-called power law (at^m) is a law widely used in the context of the extrapolation of variations of a parameter chosen as a function of time, with the aim of calculating lifetimes, but such a law is fully empirical with

10 real physical sense. However, due to the simplicity of extracting the parameters (a, m) from appropriate tools, but also because it simply adapts to a large number of parameters such as I_{th} and I_{Bias}, optical efficiency (Sauvage 2000), as we have demonstrated previously.

Even if the literature shows that the value of the parameter m is related to the degradation mode (Ikegami and Fukuda 1991), nothing really links the kinetics of degradation to the degradation mechanism itself, since parameter a has no real physical meaning.

Numerous studies have been carried out with the aim of identifying a physical model making it possible to describe the curves of parametric variations as a function of time, and in particular in the case of the I_{th} parameter, often considered as the image of the performance of a laser diode (Imai et al. 1978; Horikoshi et al. 1979; Kondo et al. 1983; Sim 1989). These studies thus make it possible to show the close relationship between the "macroscopic" parameter, and the technological parameters of the laser diode (i.e. carriers concentration generated defects rate, non-radiative traps density, etc.) (Fukuda 1986; Sobiestiankas 2005).

Among the published works, two models are particularly relevant:

– the first was proposed by Chuang et al. based on a study of the equations connecting the variations in the number of carriers (n) and the number of photons in the cavity (S);

– the second, proposed by Lam et al. takes advantage from an approach based on a population growth model suitable to mode the spread of defects within the active area.

3.6.1. Chuang's model: relationship between I_{th} and intrinsic density of defects

This model is based on the following assumption: the increase in the threshold current I_{th} of a laser as a function of time is mainly linked to an increase in the current linked to the increase in non-radiative recombinations (Chuang et al. 1997, 1998). These are related to the increase in the density of defects in the active area. Indeed, the threshold current can be described by the sum of a spontaneous emission current I_{sp} (assumed to be constant) and a non-radiative recombination current I_{nr}, which will depend on the density of defects, and is written as follows:

$$I_{th}(t) = I_{nr}(t) + I_{sp} \qquad [3.56]$$

with

$$I_{nr}(t) = \frac{qVAN_d(t)n_{th}}{\eta_i}$$

[3.57]

and

$$I_{sp} = \frac{qVBn_{th}^2}{\eta_i}$$

[3.58]

Here, V is the volume of the active area, q is the charge of the electron, A is the rate of non-radiative recombinations and B is the coefficient of radiative recombinations. $N_d(t)$ represents the rate of defects generation at time t.

Chuang *et al.* also suggest an interaction process, between defects and carriers mainly governed by the rate of defects generation, which can take the following form:

$$\frac{dN_d(t)}{dt} = K(n)N_d(t)$$

[3.59]

with

$$K(n) = \kappa n(t)p(t) = \kappa n^2$$

[3.60]

K (n) represents the coefficient depending on the physical process of defect generation. They consider that electrical carriers are balanced in the undoped are (n = p), and the annealing phenomena at the interfaces ("annealing effect") are neglected. From there, we assume that the carriers density is constant and equal to its value n_{th} at the threshold; it is fixed by the Bernard–Durrafourg's condition when the laser operates beyond its threshold. We then have $K(n_{th}) = K_{th}$ constant, and the differential equation, describing $I_{nr}(t)$, is then written as:

$$\frac{dI_{nr}(t)}{dt} = K_{th}I_{nr}(t)$$

[3.61]

After solving and putting the result in equation [3.56], we can calculate the absolute variation of I_{th} as follows:

$$\Delta I_{th}(t) = I_{nr0}\left(e^{K_{th0}te^{-\frac{E_a}{kT}}} - 1\right)$$

[3.62]

It should be noted that this model presents a significant difference with our study. It was carried out on blue-green laser diodes, that are non-mature from a technological point of view demonstrating short lifetimes in operating conditions at the end of the 1990s; a few hundred hours for II–VI materials (Zhang *et al.* 2001; Tanigashi 1996) and several hundred hours for GaN-based laser diodes (Ustinov *et al.* 2001; Nakamura 1997; De Santi 2016). In this case, the curves fitting the degradation do not follow a sigmoidal law but an exponential one.

An adjustment using this model was made on our experimental data used in section 3.3.3. Figure 3.30 represents the variations of I_{th} as a function of time for a particular device (chip 1 – see Table 3.1) as well as the theoretical curve obtained after adjustment.

In order to highlight the weakness of this model to fit the degradation of our tested devices, this theoretical curve has been extrapolated until 20% of variations were obtained, representing the failure criterion for the threshold current (Figure 3.31). However, this variation, which corresponds to a hypercritical case because it has been rarely observed for mature technologies such as DFB 1550 nm laser diodes, is obtained after 8500 h (about a year) with this model, while the failure criterion is only reached after 335,000 h of operation (almost 40 years) considering the power law.

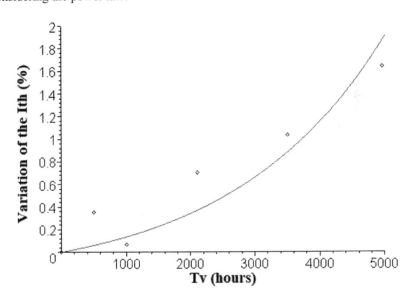

Figure 3.30. *Mathematical fitting, by Chuang's model, of the experimental data relating to the variations of I_{th} of the component E11*

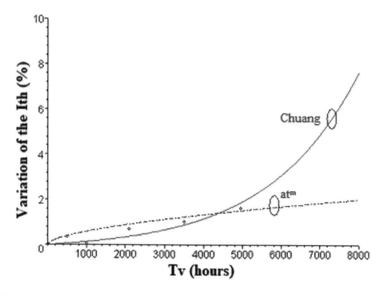

Figure 3.31. *Extrapolation of the model fitted to the data of component E11: comparison of Chuang's model (solid line) with the at^m model (dotted lines)*

This model, even though it is effective for blue laser diodes, remains difficult to apply to DFB 1550 nm laser diodes. However, if the exponential variation of the I_t parameter is not applicable on devices with proven technological maturity, the relationship between the threshold current and the fault density remains valid. Thus, an interesting contribution would benefit from the concept of saturation in the increase of defects density, agreeing the variations of threshold current observed on technologies of DFB 1550 nm laser diodes.

3.6.2. *Lam's model: growth of defects*

The model, proposed by Lam *et al.* (2003), is based on the previous approach together with a classic model of population growth. Mainly, Lam's contribution lie with the fact that the rate of defects creation or defects growth does not evolve continuously and constantly in a spatially bounded environment.

Indeed, if the variation of the threshold current is proportional to the variation of the defects concentration, then this last one must necessarily saturate at a certain time during the aging process due to the saturation of the defects growth rate. Such a hypothesis is quite correlated with the signatures of gradual and non-linear degradation over time. In addition, this model can take into account the fact that during aging, different types of defect can develop, mainly identified by their own activation energy (E_a).

As for Chuang $et\ al.$'s model, Lam's model starts from the same concept:

$$\Delta I_{th}(t) = I_{nr}(t) - I_{nr}(0) \qquad [3.63]$$

with I_{nr} defined in equation [3.57], but considers that N_d corresponds to the density of non-radiative type defects.

Taking into account the saturation of the increase in defects over time, it is necessary to write the following differential equation defining $N_d(t)$:

$$\frac{dN_d(t)}{dt} = K_n N_d(t) - C_n N_d(t)[N_d(t) - 1] \qquad [3.64]$$

where K_n and C_n are defined by:

$$K_n = \kappa np \exp\left(-\frac{E_a}{kT}\right) = \kappa n^2 \exp\left(-\frac{E_a}{kT}\right)$$

$$C_n = cnp \exp\left(-\frac{E_a}{kT}\right) = cn^2 \exp\left(-\frac{E_a}{kT}\right) \qquad [3.65]$$

K_n and C_n define the growth rate of the defects density and the saturation rate of such a density, respectively. They are both dependent on the value of n, representing the carrier density provided by the current bias during the aging. This carrier density is assumed to be constant as a function of the aging time.

The solution of the differential equation [3.64] makes it possible to obtain the variation of the threshold current in the following form:

$$\Delta I_{th}(t) = \frac{qVAn_{th}}{\eta_i}\left[\frac{M_n N_d(0)}{N_d(0) + [M_n - N_d(0)]\exp[-C_n M_n t]} - N_d(0)\right] \qquad [3.66]$$

It should be noted that this variation only considers one type of defects. However, Lam $et\ al.$ show that that the variation of I_{th} depends on several types of defects (pre-existing within the structure), each with their own kinetics of degradation, thus explaining certain nonlinearities that can appear when monitoring the drift of the threshold current over time. Thus, if we consider "m types" of defect, it will then be necessary to determine "m equations" equivalent to the ones given in [3.66] and based on a high number of parameters N_d (0), M_n, C_n and E_a.

The presented model have clearly opened a perspective towards an extrapolation law of I_{th} variations more related to the technology as well as the physics of the component than the law in at^m and directly usable in our methodology. However this model has been restricted to only one parameter (I_{th}). Therefore, it would be motivating to develop a model relating to optical efficiency (α) in order to more realistically describe the specific degradation mechanisms arising in DFB 1550 nm laser diodes.

Lam's model turns out to be complementary to the so-called power model (at^m) which is compliant with existing qualification standards in particular for high-rate telecommunication applications. Indeed, this power law can be seen as a macroscopic model making it possible to limit a physical model to only two parameters to be determined (a and m) and possibly with an application to a large variety of functional parameters of a laser diode (I_{bias}, I_{th}, V_{th}, R_S, a, Dl...). This model remains clearly more relevant for the industrial field, in particular for qualification testing, whereas Lam's model is more oriented towards R&D investigations. In conclusion, the at^m model allows rapid and practical lifetime prediction of devices from a mature technology, while the second one makes it possible to qualify a less mature technology using shorter time testing, by eliminating devices whose intrinsic defects rate (N_d) is too high. Finally, these two models are fully applicable to any technology of laser diode and emissive zone structure (i.e. edge, buried, VCSEL...) regardless of the emission wavelength.

3.7. Conclusion

Long-term reliability and extended lifetime of electronic and optoelectronic devices are a major issue in most application fields and failure rates are no longer reserved for "extreme" environments such as space or defense. This chapter describes and investigates the performance of a dedicated methodology for lifetime prediction of DFB 1550 nm laser diodes used for high-rate optical networks operating at 2.5 or 10 Gbits/s. It is important to recall that the term "very high reliability" does not necessarily mean very long lifetime, but rather that, over a given population, a very tight distribution of lifetimes is obtained around a specified "end of life" value determined by a failure criterion. In such conditions, building and demonstrating very high levels of reliability represents a scientific, technological and economic impediment. Therefore, this situation requires a complete renewal of the methods to build and demonstrate reliability since it is necessary to guarantee distributions of cumulative failures made up of little or no early failures and an almost zero failure rate in operating conditions.

This chapter details a software solution consisting of the use of statistical tools (Monte Carlo approach) to estimate the relevant parameters, from long-term truncated tests on a small population of samples. The objective is to achieve a random draw of couples of parameters with a detailed description of degradation kinetics. These couples are dependent on the performance of the component during the accelerated test and thus make it possible to create new virtual couples, thus reinforcing the statistical credibility to minimize the error of final estimation. This study has pointed out several significant results:

1) The first part of this chapter has enabled us to define the context of this study and its application. It describes the key steps of the software solution specifically developed in the Maple V environment with the collaboration of the Qualification and Reliability Department from the former ALCATEL Optronics group and based on the drawing of random numbers using a Monte Carlo approach. An application was implemented on DFB 1550 nm laser diodes of the same batch, after application of accelerated aging testing procedure (53° C-110 mA-5000 hours) with respect to TELCORDIA qualification standards. A draw on 1024 samples was performed from eight couples (a, m) selected and determined through the mathematical smoothing, by a power law, of the bias current (I_{Bias}) of eight laser diodes mounted on an AlN submount, namely the chip on submount (CoS). The maximum error between the experimental and simulated distributions is lower than 0.5% considering an initial logarithmic-type correlation law between the parameters a and m. The result obtained from five successive random draws, each performed on 1024 points, demonstrates an excellent adequacy of this technology with respect to TELCORDIA standards for this application since the failure rate, taking into account the acceleration factor (AF = 8), is estimated to be 120 FITs at 20 years with a very narrow standard deviation (6%).

2) The second part of this chapter has allowed us to assess the robustness of the method used. Firstly, a fully analytical calculation demonstrated that the Monte Carlo draw process did not introduce any specific bias into the predicted results. A crucial point was also addressed: the minimization of the estimation error of the degradation laws initially smoothed and related to each laser diode aged in accelerated conditions through the application of three types of mathematical estimator. The optimal estimator chosen is the least squares estimator. Finally, the robustness of the method was investigated through the impact of a bias on the key element of our processing chain: the initial correlation law of the parameters a and m. Thus, we have shown that even a large bias on the error Δm of the correlation law has little influence on the simulation results. Similarly, the best fitting law of the failure times is obtained from the log-normal law; Weibull's law is rejected because

of a systematic underestimation of the failure rate and atypical behavior, of the latter, especially at the beginning of the device life.

3) The last part was devoted to the possibility of discriminating failure mechanisms by taking into account a multiparametric approach, within a batch of laser diodes (20 devices) exhibiting different failure signatures. Indeed, by assuming an increase in the rate of non-radiative recombinations (see Chapters 1 and 2), that induces a reduction in optical efficiency, we have shown the advantage of reconstructing the distributions of cumulative failures taking into account variations in the threshold current (I_{th}) and the optical efficiency (α) of the laser diodes rather than the bias current (I_{Bias}). Certainly, only considering I_{Bias} can lead to extrapolating an overestimated lifetime for a fixed "end-of-life" criterion.

Clearly, these results offer interesting opportunities for this work by combining this study with the so-called "multicomponent models" (MCM) approach introduced in Chapter 2 by colleagues from McMaster University. Such an approach has clearly established the relationship between the variations of an electrical quantity (Ith) and parameters more closely related to the technology of the device (e.g. rate of native defects in the active area).

3.8. References

Bao, L., Leisher, P., Wang, J., Devito, M., Xu, D., Grimshaw, M., Dong, W., Guan, X. Zhang, S., Bai, C., Bai, J.G., Wise, D., Martinsen, R. (2011). High reliability and high performance of 9xx-nm single emitter laser diodes. *Proc. SPIE 7918, High-Power Diode Laser Technology and Applications IX*, 791806.

Béchou, L., Aupetit-Berthelemot, C., Guerin, A., Tronche, C. (2013). Performances and reliability predictions of optical data transmission links using a system simulator for aerospace applications. *34th IEEE AEROSPACE AIAA CONFERENCE, Session 11.08*, Big Sky, MT, USA.

Béchou, L., Deshayes, Y., Ousten, Y., Gilard, O., Quadri, G., How, L.S. (2015). Monte-Carlo computations for predicted degradation of photonic devices in space environment. *36th IEEE AEROSPACE AIAA CONFERENCE, Session 11.08*, Big Sky, MT, USA.

Bernstein, J.B. (2014). *Reliability Prediction from Burn-In Data Fit to Reliability Models* Academic Press, Amsterdam.

Bolam, R.J., Pamachandran, W., Coolbaugh, D., Watson, K. (2002). Electrical characteristics and reliability of UV transparent Si3N4 metal-insulator-metal (MIM) capacitors. *IEEE Transactions on Electron Devices*, 1, 1–4.

Bonfiglio, A., Casu, M.B., Magistrali, F., Maini, M., Salmini, G., Vanzi, M. (1998). A different approach to the analysis of data in life-tests of laser diodes. *Microelectronic. Reliability*, 38, 767–771.

Chuang, S.L., Ishibashi, A., Kijima, S., Nakayama, N., Ukita, M., Taniguchi, S. (1997). Kinetic model for degradation of light emitting diodes. *IEEE Journal of Quantum Electronics*, 33, 970–979.

Chuang, S.L., Nakayama, N., Ishibashi, A., Taniguchi, S., Nakano, K. (1998). Degradation of II-VI blue-green semiconductor lasers. *IEEE Journal of Quantum Electronics*, 34, 851–857.

Cui, Z., Liou, J.J., Yue, Y., Wong, H. (2005). Substrate current, gate current and lifetime prediction of deep-submicron nMOS devices. *Solid-State Electronics*, 49, 505–511.

De Santi, C., Meneghini, M., Meneghesso, G., Zanoni, E. (2016). Degradation of InGaN laser diodes caused by temperature- and current-driven diffusion processes. *Microelectronics Reliability*, 64, 623–626

Deshayes, Y., Béchou, L., Verdier, F., Tregon, B., Laffitte, D., Goudard, J.L., Hernandez, Y., Danto, Y. (2004). Estimation of lifetime distributions on 1550 nm DFB Laser diodes using Monte-Carlo statistic computations. *Proceedings Volume 5465: Reliability of Optical Fiber Components, Devices, Systems, and Networks II*, 103–115, Strasbourg, France.

Fukuda, M. (1991). *Reliability and degradation of semiconductor lasers and LEDs*. Artech House, Norwood.

Fukuda, M. and Iwane G., (1986). Correlation between degradation and device characteristic changes in InGaAsP/InP buried heterostructure lasers. *Journal of Applied Physics*, 59, 1031–1037.

Goudard, J.L., Berthier, P., Boddaert, X., Laffitte, D., Périnet, J. (2000). Reliability of optoelectronic components for telecommunications. *Microelectronics Reliability*, 40, 1701–1708.

Hammersley, J.M. and Handscomb, D.C. (1964). *Monte-Carlo Methods*. Methuen, London.

Horikoshi, Y., Koyabashi, T., Furukawa, Y. (1979). Lifetime of InGaAsP–InP and AlGaAs–GaAs DH lasers estimated by the point defect generation model. *Japanese Journal of Applied Physics*, 18, 2237.

Hwang, N., Kang, S.G., Lee, H.T., Park, S.S., Song, M.K., Pyung, K.E. (1997). An empirical lifetime projection method for laser diode degradation. *35th Annual Proceedings of the International Reliability Physics Symposium*, April 8–10, 272–275.

Ikegami, T. and Fukuda, M. (1991). Optoelectronics reliability. *Quality and Reliability Engineering International*, 7, 235–241.

Imai, H., Isozumi, K., Takusagawa, T. (1978). Deep level associated with the slow degradation of GaAlAs DH laser diodes. *Applied Physics Letters*, 33, 330–332.

Kim, K.S., Kim, H.I., Yu, C.H., Chang, E.G. (2004). Fatigue analysis of high-speed photodiode submodule by using FEM. *Microelectronics Reliability*, 44, 167–171.

Kondo, K., Ueda, O., Isozumi, S., Yamakoshi, S., Akita, K., Katoni, T. (1983). Positive feedback model of defect formation in gradually degraded GaAlAs light emitting devices. *IEEE Transaction on Electron Devices*, ED-30, 321–326.

Lam, S.K.K., Mallard, R.E., Cassidy, D.T. (2003). Analytical model for saturable aging in semiconductor lasers. *Journal of Applied Physics*, 94, 1803–1809.

Liu, X., Daykin, L., Li, Y., Boucke, K. (2017). System reliability estimation of high power diode laser with hypo-exponential distribution. *Annual Reliability and Maintainability Symposium (RAMS)*, Orlando, FL, USA.

Maliakal, A. (2019). Estimation of multimode pump ensemble reliability using Monte Carlo simulation to account for derating and variable stress profiles. *Proceedings Volume 10910, Free-Space Laser Communications XXXI*, San Francisco, CA, USA.

Mawatari, H., Fukuda, M., Matsumoto, S., Kishi, K., Itaya, Y. (1996). Reliability and degradation behaviors of semi-insulating Fe-doped InP buried heterostructure laser fabricated by MOVPE and dry etching technique. *Microelectronics Reliability*, 36 (11/12), 1915–1918.

Mendizabal, L. (2006). Fiabilité de diodes laser DFB 1,55 μm pour des applications de télécommunication : approche statistique et interaction composant-système. PhD Thesis, University of Bordeaux, France.

Mendizabal, L., Béchou, L., Deshayes, Y., Verdier, F., Danto, Y., Laffitte, D., Goudard, J.L., Houe, F. (2004). Study of influence of failure modes on lifetime distribution prediction of 1.55 μm DFB Laser diodes using weak drift of monitored parameters during ageing tests. *Microelectronics Reliability*, 44, 1337–1342.

Nakamura, S. (1997). *Proceeding of the Conference on Lasers Electro-Optics*. OSA, Baltimore MD, 318–319.

Nougier, J.P. (1987). *Méthodes de calcul numérique*, 3rd revised edition. Masson, Paris.

Park, S.W., Moon, C.K., Kang, J.H., Kim, Y.K., Hwang, E.H., Koo, B.J., Kim, D.Y., Song, J.I. (2003). Overgrowth on InP corrugations for 1.55 μm DFB LDs by reduction of carrier gas flow in LPMOCVD. *Journal of Crystal Growth*, 258, 26–33.

Resneau, P. and Krakovski, M. (2004). Long term ageing with highly stable performances of 1.55 μm DFB lasers for microwave optical links. *Reliability of Optical Fiber Components, Devices, Systems, and Networks II. Proceedings of SPIE*, 117-126.

Sauvage, D., Laffitte, D., Périnet, J., Berthier, P., Goudard, J.-L. (2000). Reliability of optoelectronics components for telecommunications. *Microelectronics Reliability*, 42 1307–1310.

Sim, S.P. (1989). A review of the reliability of III-V opto-electronic components. *Proceeding of the NATO Advanced Research Workshop on Semiconductor Device Reliability, NATO ASI Series E Applied Sciences*, 175, 301–321.

Sim, S.P. (1993). The reliability of laser diodes and laser transmitter modules. *Microelectronics Reliability*, 33(7), 1011–1030.

Sobiestiankas, R., Simmons, J.G., Letal, G., Mallard, R.E. (2005). Experimental study on the intrinsic response, optical and electrical parameters of 1.55 μm DFB BH laser diodes during aging test. *IEEE Transactions on Device and Materials Reliability*, 5, 659–664.

Suhir, E., Mahajan, R., Lucero, A., Béchou, L. (2012). Probabilistic design for reliability (PDfR) and a novel approach to qualification testing (QT). *33rd IEEE AEROSPACE AIAA CONFERENCE*, Session 11.08, Big Sky, MT, USA.

Svensson, T.K. and Karlsson, P.O.E. (2004). Deploying optical performance monitoring in TeliaSonera's network. *SPIE Photonics Europe*, 5465, 151–156.

Tanigushi, S., Hino, T., Itoh, S., Nakano, K., Nakayama, N., Ishibashi, A., Ikeda, M. (1996). 100 h II-VI blue-green laser diode. *Electronics Letters*, 32, 552–553.

Telcordia GR-468-CORE (1998). Generic reliability assurance requirements for optoelectronic devices used in telecommunications equipments. Telcordia, Issue 1.

Wang, P. and Coit, D.W. (2004). Reliability prediction based on degradation modeling for systems with multiple degradation measures. *Annual Symposium Reliability and Maintainability*, Los Angeles, CA, USA.

Zio, E. (2013). *The Monte Carlo Simulation Method for System Reliability and Risk Analysis*. Springer, London.

4

Laser Diode Characteristics

4.1. Introduction

In the past few years, several authors have proposed and developed a model for laser diodes (Vanzi 2008; Mura and Vanzi 2010; Vanzi *et al.* 2011) based on a new version of the rate equations for photons and charges.

The prompt for going back and revising the foundations of laser diode modeling themselves has been, for the authors, the difficulty of applying the available rate equations in a coherent way when analyzing DC electro-optical characteristics evolving in time, that is, when dealing with degradations. The point is the plural form "rate equations", because a single, unique form does not exist. Gain, optical power, threshold, efficiency and, in general, all quantities that are relevant for characterizing and monitoring such devices refer to different representations of injection: separation of the quasi-Fermi levels, carrier density and current. It is difficult to harmonize them and look, for instance, for a self-consistent treatment able to correlate gain saturation with current injection, or to continuously describe the transition between the ranges where spontaneous or stimulated emission dominates. An example of a difficult task is the search for a relationship between the clamp value of the quasi-Fermi levels and the threshold current, despite the fact that they are different representations of the same phenomenon: the achievement of lasing.

It is the authors' opinion that this situation occurs due to the historical evolution of laser equations, in general, and of laser diodes, in particular. For this reason a summary of the history of laser diodes may be of some help.

Chapter written by Massimo VANZI.

After Einstein's (1916, 1917) seminal papers that, in 1916–1917, first proposed the idea of stimulated radiation, for decades the research has been focused on amplification of radiation in the microwave domain (Townes *et al.* 1954; Schawlow and Townes 1958; Maiman 1960), up to the definition of a fundamental rate equation for maser given by Statz and De Mars (1960). The translation to the optical domain, moving from maser to laser, was theoretically investigated by Lamb (1964). The peculiarity of this phase was the consideration of physical systems where the probability of upward or downward energetic transitions was defined by the independent population of the excited and of the non-excited states. No mass-action laws have ever been invoked, taking into account the populations of both the initial and the final states for the same transition. This is an important point to be considered when the laser history approaches the world of solid state physics, whose milestones appeared after Einstein's works (1916, 1917). Indeed, Pauli's exclusion principle was published in 1925, the Fermi-Dirac statistics one year later (Fermi 1926; Dirac 1926) and, lastly, Bloch's theorem in 1928. This prepared the ground for the book on electrons and holes by Shockley (1950).

The two lines of research (maser/laser and solid state electronics) run nearly independently, and when they merged (Lasher 1964; Statz *et al.* 1964; Ikegami and Suematsu 1967, 1968; Paoli and Ripper 1970; Adams 1973; Salathé *et al.* 1974; Bores *et al.* 1975), the laser diode rate equations, on the one hand, looked at the formalism of the semiconductor collective states and, on the other hand, tried to harmonize themselves with the assessed heritage of maser/laser physics, developed for systems of local wave functions. It resulted in a dual description of photon-charge interactions: one, keeping the concepts of quasi-Fermi levels and band population, was mostly used for spectral properties and, in particular, for gain (including the transparency condition); the other, counting the rate of change of the number of particles, with a set of two lumped equations for charges and photons, mainly applied to current and optical power. If the former approach still shows its clear foundation on quantum mechanics, the latter generally faces the computational challenges for many quantities introducing phenomenological considerations.

Such a dual approach survived in the first fundamental books summarizing the state of the art (Kressel and Butler 1977; Case and Panish 1978; Thompson 1980) including the evolving technology that rapidly brought laser diodes to double heterostructure. Even when further studies widened the field of application of laser diodes, studying their modulation (Arnold *et al.* 1982; Henry 1985; Lau and Yariv 1985), they did not change this original dichotomy.

When we go through the currently available textbooks (Agrawal and Dutta 1986; Verdeyen 1995; Yariv 1995; Coldren *et al.* 2012), we find subjects such as spectral gain and the transparency condition in different chapters than the current–power relationship. This makes it difficult to correlate any observed kinetics with physics, as we can for degradations. As an example, if one measures an increase in the threshold current, how useful can the concept of the "nominal current" J_{nom} (Henry 1985) be, which is required to "uniformly excite a unit thickness of active material at unitary efficiency"? How can a degradation mechanism affect J_{nom}?

In a nutshell, the difference between the solid state physics and the historical rate equations for lasers is dramatically simple: the latter, neglecting the mass-action law for the non-equilibrium transitions, does not refer any relevant quantity to the voltage V. This, for a device that is built, named and largely behaves as a diode, is a serious handicap.

This chapter aims to rewrite the rate equations for a laser diode focusing on the voltage V as the main reference parameter. Nothing of laser physics is modified, but the choice is proven to greatly unify the study of the many different quantities that characterize this kind of devices.

The approach is to start with an ideal double heterostructure quantum well diode, whose inner part, the active region, is responsible for all recombination, and recombination is completely radiative (unit internal efficiency). Here, the quantum size of the active layer justifies the transversal uniformity of densities (usually assumed for infinitely extended regions) based on the loss of significance of locality on a quantum scale. The detailed band structure will be packed into the effective masses and density of states, so the simple (and widely used) parabolic band model will be employed. A specific appendix will be dedicated to provide the more defined picture that includes light and heavy hole subbands.

The k-selection rule is preserved, which implies neglecting the non-collective states in band tails. It will be shown that the basic characteristics of a real device are not affected by such approximation.

The rate equation for the ideal active region will be derived by first considering and then modifying the equilibrium state. The separation of the quasi-Fermi levels, following Paoli and Barnes (1976), is identified with the energy equivalent qV of the applied voltage V.

Several results are obtained from the solution of the rate equation, including spectral and modal gain, the ideal $I(V)$ current–voltage characteristics and the initial form of the $P_{OUT}(I)$ power–current relationship. The threshold will appear as an asymptotic value for voltage and a fast but non-abrupt transition will continuously connect the regions under and above the threshold. This bunch of results follows the simple strategy inversion in the proposed approach, when compared with the previous treatments: instead of computing the total number of electron-hole transitions and then looking for the fraction that creates light, only the radiative ones are considered, finding a harmonic, self-consistent and quite peculiar relationship between V, I and P_{OUT}, which describes the ideal diode as a device.

It is only after this step that comparison with the real world will force us to include non-radiative recombination inside the active layer itself, and to model it as an additive Shockley current, sharing the same voltage of the radiative one.

This non-radiative current at the threshold voltage will be shown to rule over the measurable threshold current. Its formulation will display the most striking difference from the literature, but it will also be shown to be, surprisingly numerically undistinguishable from the previous results, validated by experiments over the decades.

P_{OUT} (V) being available, the expression of the total (radiative and non-radiative) current I as a function of the same V will lead to the continuous L-I curve (Adams 1973; Salathé et al. 1974; Agrawal and Dutta 1986; Yariv 1995; Coldren et al. 2012), whose asymptotic limits exactly recover the well-known expression for the light emitting diode (LED) and the laser regimes.

Other currents are then considered, dominant at very low injection, in a region usually neglected in standard current-driven measurements. They will be identified and weighted for their relevance, which is null in regular devices, but can become important under degradation.

This will conclude the chapter, which is intended as the seed for further studies. A list of further non-idealities and open points will be given and discussed at the end. In that list, we include the harmonization of the new modal gain relationship with the existing ones, the puzzle of the ideality factor, the sharpness of the threshold transition and the effects of longitudinal and transversal non-uniform pumping.

With respect to the previous papers (Vanzi 2008; Mura and Vanzi 2010; Vanzi et al. 2011), the most relevant new points are:

1) the detailed derivation of the rate equation, starting from the black body case. The approach is similar to that reported in literature (Einstein 1916, 1917), but here it is specialized to the specific notation used. In particular, the joint distribution of electrons and holes is always kept in evidence as the product of two distinct densities, avoiding the step of referring rates and balances to the square or another power of a single carrier density N;

2) the inclusion and discussion of band asymmetry;

3) the discussion on the dimensionality of the joint spectral densities of electrons and holes;

4) the extension of the range of validity of the rate equation to the extremely low injection level, which results in recovering the complete Shockley-like formulation of the subthreshold regime;

5) the correlation of a previous (Vanzi 2008; Mura and Vanzi 2010; Vanzi et al. 2011) generic parameter Ω_{v0} with the gain coefficient g_0 that appears in many studies (Coldren et al. 2012) as a phenomenological term. This opens the way to a new formulation of both threshold current and gain.

Measurability being one of the goals of the present study, a method will also be recalled (Mura et al. 2013) for obtaining the separate radiative and non-radiative components I_{ph} and I_{nr} of the total current I from experimental curves, as well as the threshold and the transparency current–voltage pairs for each of them.

4.2. Energies and densities

Let us consider an ideal laser diode (Figure 4.1), where all recombination is radiative and occurs inside the very thin QW undoped active layer of a double heterostructure. The quantum size of the active layer makes any concept of density gradient meaningless, allowing electrons, holes and photons to interact irrespective of their position along the thickness of the layer itself.

This is the same as stating that the quasi-Fermi levels for electrons and holes, as well as the existing photon density, are uniform inside the active layer.

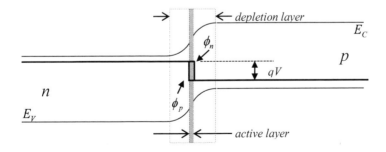

Figure 4.1. *Quasi-Fermi levels in the ideal laser diode*

Dealing with optical transitions in a semiconductor, the usual concepts (Verdeyer 1995, section 11.4) that define (Table 4.1 and Figure 4.2) electron and hole energy and density, as well as their complementary density, are introduced:

$$\begin{cases} E_e = E_C + \dfrac{\hbar^2 k_e^2}{2m_e^*} \\ E_h = E_V - \dfrac{\hbar^2 k_h^2}{2m_h^*} \end{cases} \qquad [4.1$$

$$\begin{cases} n(E_e) = \dfrac{g_e(E_e)}{\exp\left(\dfrac{E_e - \phi_n}{kT}\right) + 1} \\ p(E_h) = \dfrac{g_h(E_h)}{\exp\left(\dfrac{\phi_p - E_h}{kT}\right) + 1} \end{cases} \qquad [4.2$$

$$\begin{cases} \overline{n}(E_e) = n(E_e)\exp\left(\dfrac{E_e - \phi_n}{kT}\right) \\ \overline{p}(E_h) = p(E_h)\exp\left(\dfrac{\phi_p - E_h}{kT}\right) \end{cases} \qquad [4.3$$

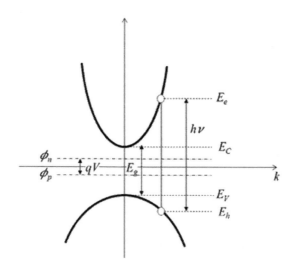

Figure 4.2. *Definition of energy levels*

E_C	Conduction band edge
E_V	Valence band edge
$E_e = E_C + \dfrac{\hbar^2 k_e^2}{2m_e^*}$	Electron energy in conduction band
$E_h = E_V - \dfrac{\hbar^2 k_h^2}{2m_h^*}$	Hole energy in valence band
$\hbar k_e$	Electron momentum in conduction band
$\hbar k_h$	Hole momentum in valence band
m_e^*	Electron effective mass
m_h^*	Hole effective mass
$g_e(E_e)\ [cm^{-3}eV^{-1}]$	Electron density of states at energy E_e
$g_h(E_h)\ [cm^{-3}eV^{-1}]$	Hole density of states at energy E_h
$n(E_e) = \dfrac{g_e(E_e)}{\exp\left(\dfrac{E_e - \phi_n}{kT}\right) + 1}\ [cm^{-3}eV^{-1}]$	Electron density at energy E_e
$p(E_h) = \dfrac{g_h(E_h)}{\exp\left(\dfrac{\phi_p - E_h}{kT}\right) + 1}\ [cm^{-3}eV^{-1}]$	Hole density at energy E_h
$\bar{n}(E_e) = n(E_e)\exp\left(\dfrac{E_e - \phi_n}{kT}\right)\ [cm^{-3}eV^{-1}]$	Complementary electron density at energy E_e

$\overline{p}(E_h) = p(E_h)\exp\left(\dfrac{\phi_p - E_h}{kT}\right)$ $[cm^{-3}eV^{-1}]$	Complementary hole density at energy E_h
ϕ_n	Quasi-Fermi level for electrons
ϕ_p	Quasi-Fermi level for holes
T (°K)	Absolute temperature
k	Boltzmann's constant
h	Planck's constant
$\hbar$	Reduced Planck's constant
q	Electron charge
ν	Photon frequency
c	Group velocity of light ($=c_0$ in a vacuum)

Table 4.1. *Definition of relevant parameters*

These elements are complemented with the quantum-mechanical selection rules on momentum and energy:

$$\begin{cases} \hbar k_e = \hbar k_h \\ E_e - E_h = h\nu \end{cases} \quad [4.4$$

The only definition that will be introduced here that is not common also relate to the difference in the quasi-Fermi levels to the electrostatic potential V.

$$\phi_n - \phi_p = qV \quad [4.5$$

This clearly recalls the similar definition as given in Shockley's (1950) theory of an ideal pn-junction, where V is the final voltage applied to the classic semiconductor diode. Here, following Paoli and Barnes (1976), it is proposed for the ideal laser diode. Physics of the two ideal devices are nearly opposite: in the Shockley diode, recombination occurs entirely outside the depletion region, while in the ideal laser diode, it is concentrated inside the active layer, fully embedded inside the depleted volume.

Let us introduce the spectral densities p_v and n_v for holes and electrons, respectively, involved in radiative transitions that create photons at frequency v. They are linked to the electron and hole densities at given energies, given in Table 4.1, by

$$n = \int_{E_C}^{\infty} n(E_e) dE_e = \int_0^{\infty} n_v dv$$

$$p = \int_{-\infty}^{E_V} p(E_h) dE_h = \int_0^{\infty} p_v dv$$

[4.6]

With such a definition, the selection rules for momentum and energy are automatically embedded in the definition of the joint density

$$p_v n_v = \int p_v n_{v_1} \delta(v_1 - v) dv_1$$

[4.7]

This is the version of the mass-action law that obeys the selection rules for a top-down transition (recombination): the probability of photon emission at energy hv is proportional to *both* the spectral density of electrons n_v and holes p_v that are tuned to frequency v. In the same way, the bottom-up transitions caused by photon absorption are linked to the complementary joint densities $\bar{p}_v \bar{n}_v$.

It is worth noticing that the dimensions are $\left[\int p_v n_v dv \right] = \ell^{-6}$, but also that $\int p_v n_v dv \neq pn$.

The listed lumped relations for densities, energies and selection rules allow us to write the joint densities that enter all processes of light emission and absorption as

$$\begin{cases} p_v n_v = \dfrac{N_v^2}{f_v^2} \\ \bar{p}_v \bar{n}_v = p_v n_v \exp\left(\dfrac{hv - qV}{kT} \right) \end{cases}$$

[4.8]

where:

$$N_v^2 = g_e\left(E_e\right)g_h\left(E_h\right)\Big|_{E_e-E_h=hv}$$ [4.9]

This is *the joint effective density of states* for an optical transition at frequency v when the k-selection rule holds, and

$$f_v^2 = \left[\exp\left(\frac{hv-qV}{kT}\right)+2\exp\left(\frac{hv-qV}{2kT}\right)\cosh\left(\varepsilon\right)+1\right]\xrightarrow[\varepsilon\approx0]{}$$
$$\xrightarrow[\varepsilon\approx0]{}\left[\exp\left(\frac{hv-qV}{2kT}\right)+1\right]^2$$ [4.10]

This is the product of the denominators of the electron and hole energy densities with

$$\varepsilon = \frac{\left(E_e+E_h\right)-\left(\phi_n+\phi_n\right)}{2kT}$$ [4.11]

The parameter ε (which we now cell the *band anomaly*) measures the asymmetry of conduction and valence bands. Referring to Figure 4.2, it should be clear that the term $\left(E_e+E_h\right)/2$ represents the energy level midway between the electron and hole energies in the given transition, while $\left(\phi_n+\phi_n\right)/2$ is the mean value of the quasi Fermi levels. Their difference is only strictly null for symmetrical bands (not necessarily parabolic), in which case both mean values are separately null. For asymmetric bands, this is not true, although it is difficult to estimate the difference at a glance. Section 4.10, appendix A, is dedicated to the discussion of the limit of equation [4.10] for vanishing values of ε. Its reading is recommended after the complete development of the model up to the definition of currents. The result, in short, is that, for practical cases, it is largely acceptable to assume $\varepsilon = 0$, which simplifies several results, and leads this treatment to agree with the original studies (Vanzi 2008; Mura and Vanzi 2010; Vanzi et al. 2011).

4.3. Rates and balances

4.3.1. Equilibrium

The most convenient starting point is the equilibrium state, when $qV = 0$. Here Einstein's century-old theory of the black-body spectrum still holds as a roadmap (Einstein 1916, 1917).

Let us add the suffix 0 to the densities of electrons and holes and introduce the photon density $\phi_{0\nu}$ at equilibrium and frequency ν.

When one specializes the Einstein approach to the black-body to the case of a semiconductor at equilibrium, the upward and downward transitions from the conduction to the valence band and the reverse process (photon absorption) give rise to rates that balance according to

$$A_{CV}\, p_{0\nu} n_{0\nu} + B_{CV}\, p_{0\nu} n_{0\nu} \phi_{0\nu} - B_{VC}\, \bar{p}_{0\nu} \bar{n}_{0\nu} \phi_{0\nu} = 0 \qquad [4.12]$$

where A_{CV}, B_{CV} and B_{VC} are coefficients, not depending on temperature, referring to photon spontaneous emission, stimulated emission and absorption.

Equation [4.8] at equilibrium gives rise to

$$\bar{p}_{0\nu} \bar{n}_{0\nu} = p_{0\nu} n_{0\nu} \exp\left(\frac{h\nu}{kT} \right) \qquad [4.13]$$

and then we have

$$A_{CV} + B_{CV} \phi_{0\nu} - B_{VC} \phi_{0\nu} \exp\left(\frac{h\nu}{kT} \right) = 0 \qquad [4.14]$$

As per Einstein's approach, for increasing temperatures, the exponential approaches unity, the photonic density increases (Stefan's law), but the coefficients remain unchanged. This leads us to identify $B_{CV} = B_{VC}$ and then to rename

$$\begin{cases} B \equiv B_{CV} = B_{VC} \\ A \equiv A_{CV} \end{cases} \qquad [4.15]$$

and then

$$\phi_{0\nu} = \frac{A}{B} \frac{1}{\exp\left(\dfrac{h\nu}{kT} \right) - 1} \qquad [4.16]$$

The usual way to proceed in the treatment of the black-body radiation will allow us to identify the ratio

$$\frac{A}{B} = \frac{8\pi \nu^2}{c^3} \qquad [4.17]$$

which gives us back Planck's formula for the power density u_ν per unit frequency

$$u_\nu \equiv \phi_{0\nu} h\nu = \frac{8\pi\nu^2}{c^3} \frac{h\nu}{\exp\left(\dfrac{h\nu}{kT}\right) - 1}$$

[4.18]

For the scopes of the current study, it will be sufficient to recognize that

$$\frac{A}{B} = \phi_{0\nu}\left[\exp\left(\frac{h\nu}{kT}\right) - 1\right]$$

[4.19]

4.3.2. Quasi-equilibrium: the rate equation

The operating condition of a real solid-state light emitter will be assumed as a steady, quasi-equilibrium state. The steady state does not assume time-dependent performances, as modulation cannot occur, but simply that, from the point of view of photons, everything happens slowly. This may seem questionable in the THz era of photonics. Focusing on light generation in semiconductor devices, we can safely assume that, in a balance among electron, hole and photon densities, the first two will have a rate of change far slower than photons.

The quasi-equilibrium, a common assumption in solid-state electronics, implies that, even under high injection level, the electron and hole distributions can be represented by a Fermi-like function, provided the common Fermi level E_F is substituted by the respective quasi-Fermi levels.

With respect to the equilibrium case given in equation [4.12], the balance must introduce the possibility that a net flux of photons enters or leaves the system. A net incoming flux will be neglected here, while an escape term is introduced. This term must be a *rate*, proportional to the photon density through a factor with the dimensions of the reciprocal of a time that assumes the meaning of the mean permanence time τ_C. This net escape rate must vanish at equilibrium, and then the *rate equation for the steady state* reads as

$$\frac{\partial \phi_\nu}{\partial t} = \underbrace{Ap_\nu n_\nu}_{R_{sp}} + \underbrace{Bp_\nu n_\nu \phi_\nu}_{R_{st}} - \underbrace{Bp_\nu n_\nu \exp\left(\frac{h\nu - qV}{kT}\right)\phi_\nu}_{R_{abs}} - \underbrace{\frac{\phi_\nu - \phi_{0\nu}}{\tau_C}}_{R_{loss}} = 0$$

[4.20]

where the specific rates have been indicated:

- spontaneous photon emission (electron–hole recombination) rate R_{sp};
- stimulated photon emission (electron–hole recombination) rate R_{st};
- photon absorption (electron–hole pair generation) rate R_{abs};
- photon loss (escape and internal losses non-generating e-h pairs) rate R_{loss}.

Even before solving equation [4.20] for the photon density ϕ_V, some general considerations are as follows:

1) The ratio R_{st}/R_{sp} between the stimulated and the spontaneous emission states that the former rapidly increases with photon density, which legitimates neglecting the spontaneous term at high injection in many studies. It is the spontaneous emission that provides the stimulus for the stimulated one, and also, mathematically, this term will be crucial for achieving the solution.

$$\frac{R_{st}}{R_{sp}} = \frac{\phi_V}{A/B} = \frac{\phi_V}{\phi_{0V}\left[\exp\left(\dfrac{h\nu}{kT}\right) - 1\right]} \qquad [4.21]$$

2) The ratio R_{st}/R_{abs} between stimulated emission and absorption shows that transparency (the situation for which the two phenomena exactly balance) occurs exactly at

$$qV = h\nu \qquad [4.22]$$

This means that each light frequency will reach transparency at a different injection level.

3) The loss rate R_{loss}, once summed over the frequency range and multiplied by the volume Vol of the active material, counts the total number of photons lost per second. If the same is calculated after multiplying R_{loss} by the photon energy $h\nu$, one gets the total photon energy leaving the active region per second, that is, the total optical power P_{TOT} emitted by the device. This is not directly the measured output power P_{TOT}, but the latter is a fraction of it.

$$P_{TOT} = Vol \int_0^\infty h\nu \cdot R_{loss} d\nu \qquad [4.23]$$

4) In a similar way, the sum of the first three terms is the *net number of radiative recombination* events at frequency ν. This, once multiplied by the electron charge q and the volume Vol, and summed over all frequencies, gives a current I_{ph}. In our

ideal laser diode, this would be the total current flowing through the device. In a real one, it is the fraction of the total current I that is in charge of sustaining the total light emission.

$$I_{ph} = Vol \cdot q \int_0^\infty \left(R_{sp} + R_{st} - R_{abs} \right) dv \qquad [4.24]$$

Before solving the rate equation, and making all the equations above explicit, it is highly convenient to introduce the coefficients

$$g_m = \frac{BN_v^2}{c} \qquad [4.25]$$

$$\alpha_T = \frac{1}{c\tau_C} \qquad [4.26]$$

where c is the group velocity of light, B is the Einstein coefficient common to stimulated emission and absorption, N_v^2 is the joint effective density of states as defined in equation [4.9] and τ_C is the photon permanence time introduced in equation [4.20].

4.3.2.1. Physical meaning of g_m

Three quantities merge into the coefficient g_m: the Einstein coefficient for stimulated transitions B, the group velocity of light c, inside the active material, and N_v^2, the joint density of states for electrons and holes involved in direct transition at energy hv.

It is then a term entirely related to *material* properties, no matter the shape of the optical cavity. In other words, g_m does not depend on losses. It is a spectral function and also a measurable quantity. It will be shown to coincide with the lower (negative) limit for the optical gain, that is, g_m represents the *absorption coefficient for the un-pumped material*.

g_m plays a crucial role, together with the loss coefficient α_T, for defining the threshold conditions for the laser regime.

Regarding the spectral properties, even if we can assume B and c as slowly varying functions of the frequency v, the joint density of states N_v^2 carries the information of the dramatic transition between forbidden and allowed states. For a ideal QW, the expected step function, centered at $hv = E_g$ (E_g being the amplitude of the band gap), suitable for taking into account some broadening, no matter its origin, will be shown to properly reconstruct the main features of a real spectrum.

4.3.2.2. Physical meaning of α_T

The coefficient α_T is simple in itself: it includes all mechanisms leading to photon losses, with the exception of the sole photon absorption which creates an electron–hole pair, and is separately taken into account in the gain term. Such "other" mechanisms can be divided as follows:

– internal losses, such as the ones caused by free electrons, for instance, or by diffused lattice defects, and that, in our model, can be considered uniformly distributed across the whole optical cavity, and described by the internal loss coefficient α_i;

– escape losses, which describe the fraction of photons which, during their travel towards the boundaries of the optical cavity, leave of the cavity itself, and do not contribute more to the photon–carrier interactions inside the active material. They should and could be indicated by a specific coefficient α_{esc} which, summed to α_i, gives α_T.

The difficulty arises from the customary tendancy to deal with escape losses in quite different ways: loss coefficients for longitudinal losses, a confinement factor for transversal losses and a variety of solutions (including no solution) for lateral losses. Sections 4.11 and 4.12, appendices B and C, will deal with losses and justify the choice to really consider α_T the sole loss descriptor.

With such coefficients in equations [4.25] and [4.26], the steady-state equation [4.20], after dividing by c and using equation [4.8] for the joint electron–hole densities, becomes a *spatial rate of change* equation:

$$\frac{\partial \phi_v}{\partial(ct)} = g_m \frac{\left[\exp\left(\frac{hv}{kT}-1\right)\right]}{f_v^2}\phi_{0v} + g_m \frac{\left[1-\exp\left(\frac{hv-qV}{kT}\right)\right]}{f_v^2}\phi_v - \left[\alpha_T\left(\phi_v-\phi_{0v}\right)\right]=0 \quad [4.27]$$

It can be now be solved for the excess photon density $\phi_v - \phi_{0v}$, also making f_v^2 explicit by means of equation [4.10]:

$$\phi_v - \phi_{0v} = \phi_{0v} \frac{\exp\left(\frac{qV}{kT}\right)-1}{\frac{\alpha_T}{g_m}\left[1+\exp\left(\frac{qV-hv}{2kT}\right)\right]^2 + \left[1-\exp\left(\frac{qV-hv}{kT}\right)\right]} \quad [4.28]$$

In parallel with equation [4.28], which is the cardinal point of the whole model let us give an explicit form to *spectral gain g*, which is defined as

$$g \equiv \frac{R_{st} - R_{abs}}{v_g \phi_v} = g_m \frac{\left[1 - \exp\left(\dfrac{h\nu - qV}{kT} \right) \right]}{\left[1 + \exp\left(\dfrac{h\nu - qV}{2kT} \right) \right]^2} \qquad [4.29]$$

Of course, both equations [4.28] and [4.29] are worth extended discussion.

4.4. Photon density

It is interesting to observe that all device or material-specific issues in equation [4.28] are given as a ratio of $\dfrac{\alpha_T}{g_m}$. This ratio plays an important role, mostly because of the behavior of g_m. It is evident that when g_m is null, the whole right-hand side of the equation is null, and $\phi_v = \phi_{0v}$. This means that only photons with energies $h\nu$ close to, or higher than, the amplitude of the energy gap E_g will contribute to the excess density $\phi_v - \phi_{0v}$.

Moreover, this also means that for values of the applied voltage qV lower than E_g by some kT, the exponentials in the denominators are negligible with respect to the unity:

$$\phi_v - \phi_{0v} \xrightarrow[qV < E_g]{} \phi_{0v} \frac{\exp\left(\dfrac{qV}{kT} \right) - 1}{\dfrac{\alpha_T}{g_m} + 1} \qquad [4.30]$$

This voltage range represents the subthreshold regime for a laser diode, where the spontaneous emission rate dominates over the stimulated emission phenomena by far. It is a range where the laser behaves as an LED, and equation [4.30] can accordingly be assumed as the representation of an LED spectrum. It includes all the coarse features of that spectrum: peaking and broadening. It does not include the fine structure due to resonances, because equation [4.20] counts particles and does not deal with waves.

In Figure 4.3, the subthreshold spectrum of a real laser diode emitting at 950 nm (black solid line) and the spectrum calculated by means of equation [4.28] (red solid line) are plotted together. The vertical scale has been suitably adjusted for good superposition, while a smoothed step-function has been used for g_m (dashed blue line), keeping α_T constant. The thin, nearly vertical dashed black line represents the term ϕ_{0v} as given by the black-body theory.

Despite the many approximations, which will be removed when measuring gain, the overlap of experimental and calculated curves, at least at this coarse level of spectral detail, looks fine. g_m and the state density and the band gap rule over the low-energy side of the peak, while the black-body function ϕ_{0v} bounds the high-energy side of the spectrum.

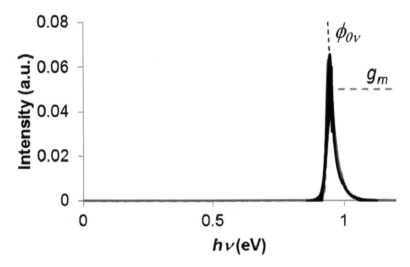

Figure 4.3. *A real subthreshold spectrum (black solid line) and the spectrum calculated by means of equation [4.9] (red solid line). For a color version of this figure, see www.iste.co.uk/vanzi/reliability.zip*

Until equation [4.30] approximates equation [4.28], any change in the injection level, by changing the applied voltage V, will result in vertically scaling up or down the same plot as in Figure 4.3.

The feature that is by far the most peculiar of equation [4.28] comes out when, fixing a photon energy $h\nu$ in the range where g_m is not null, we plot it as a function of the applied voltage V. We discover that there does exist a possible critical value V_{th} that leads the denominator to vanish:

$$qV_{th} = h\nu + 2kT \ln\left(\frac{1 + \dfrac{\alpha_T}{g_m}}{1 - \dfrac{\alpha_T}{g_m}}\right)$$

[4.31]

This means that, fixing a specific photon energy $h\nu$, and plotting the increase of the photon density beyond its equilibrium value as a function of the voltage V, one has a curve as in Figure 4.4, whose solutions for $V > V_{th}$ have been neglected because a real device should be required first to reach infinite photon density, and then infinite energy. A better physical insight comes from looking at Figure 4.4 as a plot of the voltage V as a function of the photon densities, that is, by interchanging the vertical and the horizontal axes. It is the first evidence for the voltage clamp, which is one of the most striking peculiarities of a laser diode.

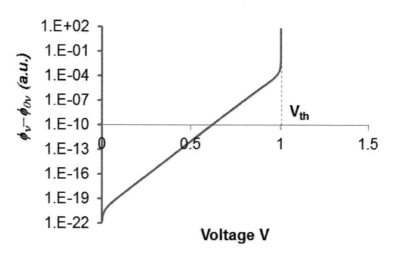

Figure 4.4. Plot of the excess photon density $\phi_V - \phi_{0V}$ as a function of the applied voltage V, according to equation [4.8], for an ideal device emitting photons at energy close to 1 eV

It is important to realize that Figure 4.4 has been calculated for a specific photon energy, and that, by means of equation [4.31], each energy has a corresponding threshold voltage V_{th}. This states that the actual threshold voltage of the whole device will correspond to the photon energy for which equation [4.31] gives the minimum value for qV_{th}.

Figure 4.5 is analogous to Figure 4.3, that is, it draws the spectrum calculated by means of equation [4.28] for several values of the applied voltage V in an ideal device whose active region is a quantum well of material with its band gap exactly at $E_g = 1$ eV, with both some broadening and a sharp step function for the density of states, and a constant ratio $\alpha_T / g_m = 0.6$ for $h\nu > E_g$. It has been drawn with a logarithmic vertical scale in order to appreciate the whole range of values in a single spectrum and also in comparison with the other.

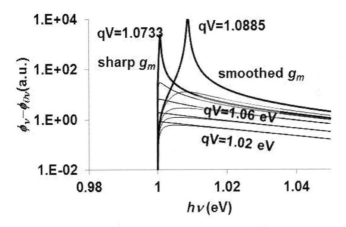

Figure 4.5. *Spectral plot of the excess photon density* $\phi_\nu - \phi_{0\nu}$, *in logarithmic scale, for different values of the applied voltage V, and for a sharp or smoothed step function for g_m, according to equation [4.8], for an ideal device with a band gap at 1 eV*

The curves from 1.02 to 1.06 eV look identical, apart from a vertical shift. Lower voltages will simply replicate them shifted downward. This represents the case in which the approximation of equation [4.28], by means of equation [4.30] holds and corresponds to the logarithmic representation of Figure 4.3. As the voltage approaches a critical value, that is, for our case, 1.0834 V, a single peak rises and grows unlimitedly (remember the log scale in Figure 4.5). Its coordinate on the horizontal axis exactly corresponds to the photon energy for which equation [4.31] gives a minimum value for V_{th}. If g_m were a sharp step function and α_T a constant, the peak would be found exactly at the leftmost side of the spectrum, that is, at $h\nu_{peak} = E_g$.

4.5. Spectral gain

Going back to equations [4.27] and [4.29], we have the definition for spectral optical gain g as:

$$g = g_m \frac{1 - e^{\frac{hv - qV}{kT}}}{f_v^2} \qquad [4.32]$$

Apart from the use of different symbols, equation [4.32] is completely identical to a popular expression available in the literature (Verdeyen 1995; Vanzi *et al.* 2016; Agrawal and Dutta 1986; Case and Panish 1978; Lasher and Stern 1964). Its explicit form is given in equation [4.29], which can be further simplified as

$$g = g_m \frac{1 - e^{\frac{hv - qV}{2kT}}}{1 + e^{\frac{hv - qV}{2kT}}} \qquad [4.33]$$

and accordingly plotted, keeping in mind that g_m is also a *spectral* function (a smoothed step centered at $hv = E_g$), depending on frequency v but not on voltage V.

It looks mathematically evident that g coincides with $-g_m$ for low V and with $+g_m$ for high V (Figure 4.6)

$$-g_m \le g \le g_m \qquad [4.34]$$

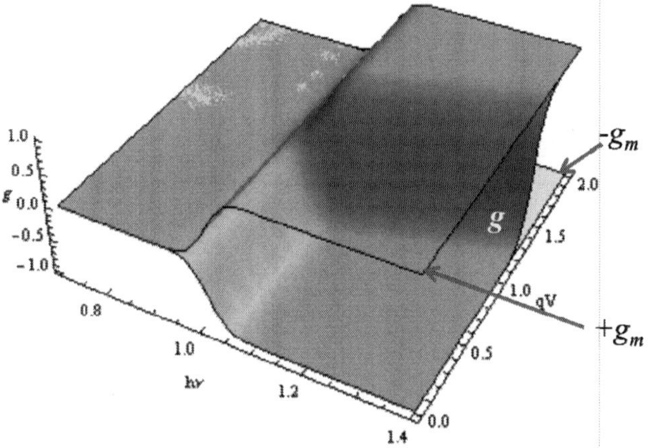

Figure 4.6. *The function $g(hv, qV)$ for a suitably smoothed step-function $g_m(hv)$. For a color version of this figure, see www.iste.co.uk/vanzi/reliability.zip*

A better insight is given by the 2D plot of the same function, i.e. by setting fixed values of qV (Figure 4.7).

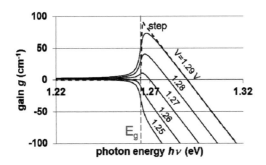

Figure 4.7. *Calculated spectral gain g at various bias V, for an active material with E_g=1.265 eV. Continuous curves correspond to Lorentzian-smoothed values of the gain coefficient g_m. The dashed line, plotted for comparison, draws the calculated gain at the highest V for a sharp step g_m. For a color version of this figure, see www.iste.co.uk/vanzi/reliability.zip*

Several interesting features shown in Figures 4.6 and 4.7 are as follows:

– gain values are negative for the majority of the values of V:

> *Going back to the foundation of the constitutive elements of the rate equation, from which gain itself springs out, this means that photon absorption usually dominates over stimulated photon emission. The separation of the quasi-Fermi levels (and then qV, because of equation [4.5]) must reach and even exceed the amplitude E_g of the band gap before having positive values for gain g;*

– the condition $g = 0$ represents *transparency*, that is, a very peculiar situation which perfectly balances photon-stimulated emission and absorption;

– the spectral range of the possible positive gain values is strictly confined close to the band gap energy.

What Figures 4.6 and 4.7 do not show is gain saturation.

It is sufficient to invert equation [4.33] to get

$$qV = hv + 2kT \ln \left(\frac{1 + \dfrac{g}{g_m}}{1 - \dfrac{g}{g_m}} \right) \qquad [4.35]$$

and to compare it with equation [4.10] to realize that the latter is the upper limit of equation [4.31], and that such limit corresponds to the *threshold condition*

$$g_{MAX} = \alpha_T \qquad [4.36]$$

Equation [4.36] may be merged into equation [4.34] for a more general form for gain in a laser diode (or better, in a laser in general)

$$-g_m \leq g \leq \alpha_T < g_m \qquad [4.37]$$

This describes the range of possible gain values, and states that if a device has $\alpha_T > g_m$, it will never reach the threshold, because of the mathematical upper bound of equation [4.37].

This points out another not-so-intuitive feature of gain, in addition to the possibility of negative gain previously pointed out: photon emission increases with increasing gain, but beyond the laser threshold any further injection will cause *more emission, but no more gain*.

The graphic representation of this relationship is shown in Figure 4.8, where the plots of g with respect to the internal voltage V and the external voltage V_{ext} are compared to the mathematical curve of equation [4.33] at a fixed photon energy.

The external voltage is the experimentally measurable quantity, which will be shown, in Chapter 5, to be easily linked to the voltage V, driving the transition rates, by adding a simple ohmic contribution. This will be made clear after introducing the total current I. What is interesting is that, with respect to the external measurement, gain saturates.

There is a beautiful mathematical form that directly links the photon density ϕ_ν and the gain function g. We first consider the expression in equation [4.28] at transparency, that is when $qV = h\nu$, and consequently (equation [4.33]) $g = 0$.

$$\left(\phi_\nu - \phi_{0\nu}\right)_{transparency} = \frac{A}{B}\frac{g_m}{4\alpha_T} \approx \phi_{0\nu}\frac{\exp\left(\dfrac{h\nu}{kT}\right)}{4\dfrac{\alpha_T}{g_m}} \qquad [4.38]$$

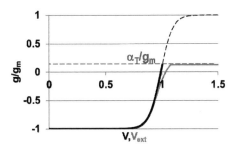

Figure 4.8. *Relative gain g/g$_m$ for a device for which $\alpha_T < g_m$. The dashed curve represents the mathematical plot of equation [4.33] as a function of the internal voltage qV. The thick solid black line is the same after introducing the voltage clamp of equation [4.31]. The thick blue line is the same as the black solid line, plotted versus the external voltage V$_{ext}$. For a color version of this figure, see www.iste.co.uk/vanzi/reliability.zip*

The unity in the numerator has been neglected because of the dominance of the exponential term for any photon energy $h\nu$ in a real spectrum.

Now, we eliminate qV from equation [4.28] by means of equation [4.35], again neglecting the unity in the numerator:

$$\varphi_v - \varphi_{0v} = \left(\varphi_v - \varphi_{0v}\right)_{transparency} \frac{\left(1 + \dfrac{g}{g_m}\right)^2}{1 - \dfrac{g}{\alpha_T}} = \frac{A}{4B}\left(1 + \frac{g}{g_m}\right)^2 \frac{g_m}{\alpha_T - g} \quad [4.39]$$

Recalling that g ranges between $-g_m \leq g \leq g_m$, as stated in equation [4.34], we see that:

– for $g = -g_m$, the limit of the un-pumped material, we only have the thermal emission at equilibrium $\varphi_v = \varphi_{0v}$;

– for $g = 0$, we recover the transparency value;

– if $\alpha_T > g_m$, the denominator never vanishes, the device never achieves the laser regime and the photon density saturates, for $g \to g_m$, at $\varphi_v - \varphi_{0v} =$

$$\left(\varphi_v - \varphi_{0v}\right)_{transparency} \frac{4}{1 - \dfrac{g_m}{\alpha_T}} = \varphi_{0v} \frac{\exp\left(\dfrac{h\nu}{kT}\right)}{\dfrac{\alpha_T}{g_m} - 1} \; ;$$

– if $\alpha_T < g_m$, the photon density is allowed to increase unlimitedly as $g \rightarrow \alpha_T$.

The construction of the gain function must be completed with the consideration that the same voltage V drives the light emission at all frequencies.

This can be visualized (Figure 4.9) by superimposing a new surface, that is $\alpha_T(h\nu)$, to the gain plot in Figure 4.6. The intersection between $g(h\nu, qV)$ and $\alpha_T(h\nu)$ defines a curve $qV_{TH}(h\nu)$ in the plane $(h\nu, qV)$, whose *minimum* has the coordinate pair $(h\nu_{peak}, qV_{th})$ which gives the laser emission energy $h\nu_{peak}$ and the device threshold voltage V_{th}.

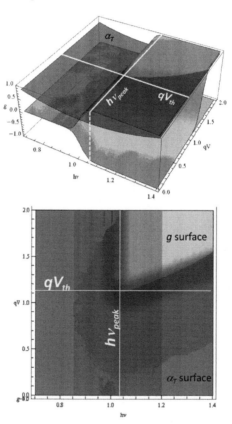

Figure 4.9. *Perspective (top) and ortographic top view (bottom) of the intersection of the gain $g(h\nu, qV)$ and loss $\alpha_T(h\nu)$ that define both the threshold voltage V_{th} and the peak emission energy $h\nu_{peak}$. For a color version of this figure, see www.iste.co.uk/vanzi/reliability.zip*

The ultimate aspect of the allowed gain curves, taking losses into account, is shown in Figures 4.10 and 4.11.

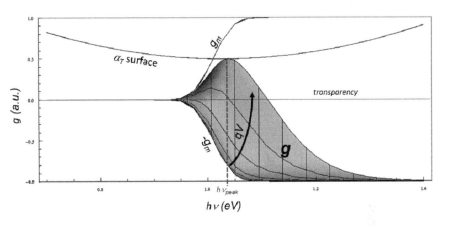

Figure 4.10. *Allowed values for spectral gain* $g_{qV}(hv)$. *For a color version of this figure, see www.iste.co.uk/vanzi/reliability.zip*

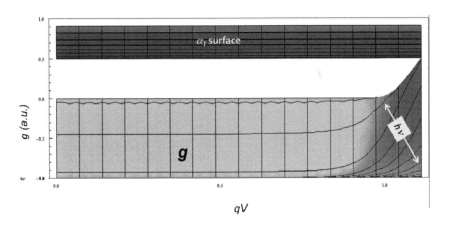

Figure 4.11. *Allowed values for modal gain* $g_{hv}(qV)$. *For a color version of this figure, see www.iste.co.uk/vanzi/reliability.zip*

It should be evident from Figures 4.10 and 4.11 that at the threshold, the perfect balance $g = \alpha_T$ only holds for hv_{peak}. All other emission lines remain strictly at $g < \alpha_T$. Once again, from this threshold point, any further current injection will only add photon emission to $h\nu_{peak}$. Gain g will remain blocked for all energies in the spectrum, and the internal voltage will stop at qV_{th}.

4.6. Integral quantities: current I_{ph} and total power P_{OUT}

It is now possible to deal with the integral quantities defined by equations [4.23] and [4.24], which are, respectively, the total emitted power P_{TOT} and the current I_{ph} that is totally converted into light and that, for the moment, is the sole current flowing across our device.

If we refer to a real spectrum, as in Figure 4.3, we see that its non-negligible values are confined within a narrow region in which the value of $h\nu$ does not change too much. In the given example, it ranges from 0.9 to 1 eV. This allows us to assume its value at the spectrum peak $h\nu_{peak}$ as a constant in equation [4.23], and write referring to equation [4.20] for an explicit form of R_{loss}:

$$P_{TOT} = Vol \cdot h\nu_{peak} \int_0^\infty \frac{\phi_\nu - \phi_{0\nu}}{\tau_C} d\nu \qquad [4.40]$$

Now, in the steady state, equation [4.20] states that $R_{loss} = R_{sp} + R_{st} - R_{abs}$, so that equation [4.24] becomes

$$I_{ph} = Vol \cdot q \int_0^\infty \frac{\phi_\nu - \phi_{0\nu}}{\tau_C} d\nu \qquad [4.41]$$

It follows that current and power are simply *proportional*, at least for the ideal case, at *any injection level*.

$$P_{TOT} = \frac{h\nu_{peak}}{q} I_{ph} \qquad [4.42]$$

The explicit calculation of the integral in equations [4.40] and [4.41] is complicated in general by the spectral behavior of the ratio $\dfrac{\alpha_T}{g_m}$, which enters the definition of $\phi_\nu - \phi_{0\nu}$, equation [4.28], and cannot be neglected. It is also complicated by the frequency dependence of τ_C.

It can be easily demonstrated (section 4.13, appendix D) that an excellent approximation of the integrals is given by

$$\int_0^\infty \frac{\phi_\nu - \phi_{0\nu}}{\tau_C} d\nu \approx \frac{kT}{h} \left. \frac{\phi_\nu - \phi_{0\nu}}{\tau_C} \right|_{\nu_{peak}}$$ [4.43]

that is, the same spectral function given by equation [4.28], calculated at the peak frequency ν_{peak} and multiplied by $\dfrac{kT}{h}$, also represents the integral function that describes the radiative current I_{ph} and the total emitted optical power P_{TOT}.

The current I_{ph} , equation [4.41], can then be accordingly expressed as

$$I_{ph} \approx Vol \cdot q \frac{kT}{h} c\alpha_T \left. \left(\phi_\nu - \phi_{0\nu}\right) \right|_{\nu_{peak}}$$
$$= Vol \cdot q \frac{kT}{h} c\alpha_T \phi_{0\nu} \frac{\exp\left(\dfrac{qV}{kT}\right) - 1}{\dfrac{\alpha_T}{g_m}\left[1 + \exp\left(\dfrac{qV - h\nu}{2kT}\right)\right]^2 + \left[1 - \exp\left(\dfrac{qV - h\nu}{kT}\right)\right]}$$ [4.44]

On a practical ground, this can be summarized as

$$I_{ph} = \begin{cases} I_{ph0}\left(e^{\frac{qV}{kT}} - 1\right) & V < V_{th} \\[2ex] \infty & V = V_{th} \end{cases}$$

where $I_{ph0} = \left[Vol \cdot q \dfrac{kT}{h} c\phi_{0\nu\,peak}\right] \dfrac{\alpha_T g_m}{\alpha_T + g_m}$ [4.45]

Here, the term I_{ph0} plays the role of a saturation current, as for a standard Shockley diode.

The plot of $I_{ph}(V)$ is then the same as in Figure 4.4, calculated at the peak frequency.

An explicit analytic formulation of the bi-modal behavior of I_{ph} as V approaches V_{th}, more refined than the rough approximation in equation [4.45], is given by the inversion of the $I_{ph}(V)$ relationship. Let us indeed consider values of qV large enough to make $\exp\left(\dfrac{qV}{kT}\right) \gg 1$ in the numerator of equation [4.44] and also let us safely assume $\exp\left(\dfrac{h\nu_{peak}}{kT}\right) \gg 1$ in the definition of ϕ_{0vpeak} in the same equation.

Some simple steps lead to:

$$qV = h\nu + 2kT \ln\left(\frac{1 + \dfrac{\alpha_T}{g_m}}{\sqrt{1 + \dfrac{I_0}{I_{ph}}\left(1 + \dfrac{\alpha_T}{g_m}\right)} - \dfrac{\alpha_T}{g_m}} \right)$$

where $I_0 = I_{ph0}\exp\left(\dfrac{h\nu_{peak}}{kT}\right)$ [4.46]

This result should be compared with both equation [4.31] and equation [4.35] which separately link the voltage V to the variable gain g and its upper value V_{th} to the maximum allowable gain $g = \alpha_T$. Here, V spans the whole range of possible values for the forward bias, and asymptotically approaches its limit for increasing current I_{ph}.

It is clear indeed that the continuous function in equation [4.46] leads to the value V_{th} as I_{ph} is allowed to increase.

The new further parameter I_0, here introduced for the sole sake of graphic simplicity, nevertheless brings a physical meaning: it is the value of the current I_{ph} at transparency for the peak emission $qV = h\nu_{peak}$

From the electrical point of view only, equations [4.44]–[4.46] describe the current–voltage relationship of a *bimodal* device that operates *like* a conventional diode in parallel with a voltage regulator, such as a Zener diode, trimmed at $V = V_{th}$. This image of two separate diodes is misleading about the delicate balance of peculiar recombination mechanisms that occur inside the same device, and so we prefer to propose a specific electronic symbol (Figure 4.12), where L and D stand for Laser Diode, for the ideal device that completely converts its current into light.

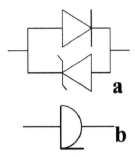

Figure 4.12. *(a) The equivalent circuit for a laser diode and (b) a proposal for a new symbol for the ideal electronic element described by $I_{ph}(V)$*

4.7. Non-radiative current I_{nr}, threshold current I_{th} and the light–current curve

Even considering that the actually measured optical power, P_{OUT}, is a fraction of the total power, P_{TOT}, the experimental evidence tells us that equation [4.42] is simply wrong. Figure 4.13 shows a measured $P_{OUT}(I)$ together with the expected curve, which should be a straight line from the origin.

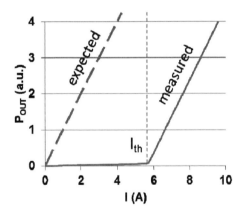

Figure 4.13. *Expected and measured emitted optical power P_{OUT} as a function of the injected current I*

The dramatic difference is the sharp change in the measured slope at a critical value that is then identified as the threshold current I_{th}. This means that the amount of current converted into light suddenly changes at $I = I_{th}$.

4.7.1. *Non-radiative current I_{nr} and the total current I*

The explanation is as simple as it is crucial: at least for $I < I_{th}$, another current I_{nr} exists, which represents *non-radiative* recombination.

What is interesting is that we can continue thinking of our ideal device, as illustrated in Figure 4.1, where *all recombination* takes place inside the active region. In this case, that *same difference* in the quasi-Fermi levels that rules over photon emission rate also drives the non-radiative recombination. This means that we can assign a Shockley equation to the non-radiative current I_{nr}

$$I_{nr}(V) = I_{nr0}\left[\exp\left(\frac{qV}{kT}\right) - 1\right]$$ [4.47]

where V is the same as in equation [4.28] for the photon density $\phi_V - \phi_{0V}$, which is the root for all other quantities, and I_{nr0} plays the standard role of a saturation current as in Shockley's theory (1950).

In order to explain the experimental evidence, we must assume that I_{nr0} in equation [4.47] is much larger than I_{ph0} in equation [4.45], and realize that a Shockley current such as I_{nr} is ruled by recombination mechanisms that do not include stimulated emission. In other words, equation [4.47] holds for any applied voltage V

When we consider the total current I, due to the competing recombination mechanisms that take place inside the active region, we must only add the two contributions I_{nr} and I_{ph}, and recognize that V is the common voltage for all

$$I(V) = I_{nr}(V) + I_{ph}(V)$$ [4.48]

It is also straightforward to introduce this element in graphical form, adding a standard diode, representing I_{nr}, to the ideal laser diode, that leads the sole current I_{ph} and is represented by the new symbol in Figure 4.12(b).

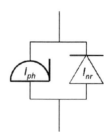

Figure 4.14. *A new equivalent circuit to represent both currents I_{ph} and I_{nr} that jointly account for all recombination in the active region of a laser diode*

In order to explain Figure 4.13, we must assume that, for $V<V_{th}$, the non-radiative current I_{nr} is, by far, dominating over the radiative component I_{ph}, that is, the quantum efficiency η_q, defined for $V < V_{th}$ as

$$\eta_q = \frac{I_{ph}}{I} = \frac{I_{ph0}}{I_{nr0} + I_{ph0}}$$ [4.49]

is a number much smaller than unity.

This is described in Figure 4.15 where the thin solid line and the thin dashed line parallel to it draw, respectively, I_{ph} and I_{nr}, and the semitransparent thick line represents the total current I. The kink in the total current has V_{th}, as its abscissa, and individuates the corresponding current value I_{th}, which is one of, if not possibly the most popular parameter in the electrical characterization of a laser diode.

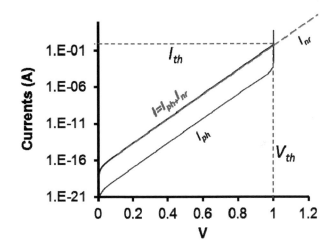

Figure 4.15. *The total current I and its components I_{ph} and I_{nr}. The threshold current I_{th} is then defined as $I(V_{th})$. For a color version of this figure, see www.iste.co.uk/vanzi/reliability.zip*

The dashed continuation of the I_{nr} line is only to indicate how that current *would* continue to increase if the voltage could exceed the threshold value V_{th}. In reality, this is exactly what cannot occur: because of the voltage clamp, the current I_{nr} is forced to stop at $I_{nr}(V_{th})$. On the other hand, the last value for the non-radiative current is also undistinguishable from the value of the threshold current I_{th}, as shown in Figure 4.15.

For this reason, and once again neglecting the unity in the Shockley for equation [4.47], we can here define the proper expression for the threshold current as:

$$I_{th} = I_{nr}\left(V_{th}\right) = I_{th0}\left[\frac{1 + \dfrac{\alpha_T}{g_m}}{1 - \dfrac{\alpha_T}{g_m}}\right]^2$$

where $I_{th0} = I_{nr0}\exp\left(\dfrac{h\nu_{peak}}{kT}\right)$ [4.50]

The relationship between I_{th0} and I_{nr} is the same as we saw in equation [4.46] for the constant I_0 and the radiative current I_{ph}: the 0-indexed term is the value of the related current at transparency, that is, at $qV = h\nu_{peak}$. But here the new constant I_{th} assumes even more significance: it is the theoretical minimum value that the important parameter I_{th} would reach in case of vanishing total losses, $\alpha_T \to 0$. It is then also a way for considering that transparency and threshold only differ because of losses.

One numerical peculiarity of equation [4.50] is worth noticing: the term in square brackets mimics an exponential surprisingly well. This means that an experiment on a real laser diode, in which the threshold current is monitored during a controlled variation of the optical losses, would measure a relationship undistinguishable from

$$I_{th} = I_{nr}\left(V_{th}\right) = I_{th0}\exp\left(4\frac{\alpha_T}{g_m}\right)$$ [4.51]

This point has been extensively discussed and experimentally and mathematically investigated in Vanzi et al. (2013), which compares various gain models and their representation of the threshold condition.

The interesting point is that a phenomenological parameter g_0, which enters some popular empirical exponential models in the literature (Coldren et al. 2012) is here identified in $g_0 = g_m/4$, and then related to the sound physical parameter that represents the optical absorption properties of the un-pumped material of the active region. This point is discussed more extensively in the next section.

4.7.2. *Gain–current relationship*

The completion of the relevant currents allows for a nice representation of the gain–current relationship which includes all the relevant features of such a link.

It is sufficient to use equation [4.35] to replace the internal voltage V with the gain g in equation [4.48] to obtain (neglecting the unity in the Shockley equations)

$$I = I_{th0} \left[\frac{1 + \dfrac{g}{g_m}}{1 - \dfrac{g}{g_m}} \right]^2 + I_0 \frac{\left[1 + \dfrac{g}{g_m} \right]^2}{1 - \dfrac{g}{\alpha_T}} \qquad [4.52]$$

This is an inverse relationship that expresses the total current I as a function of the gain g, and that cannot be inverted in a simple form. It allows us to plot the $g(I)$ curve and clearly displays several important features as follows:

1) gain has a minimum when $I = 0$, which corresponds to $g = -g_m$;

2) at transparency, when gain vanishes, by definition, the current is $I = I_{th0} + I_0$, in agreement with the definition of the constants I_0 and I_{th0} in equations [4.46] and [4.50], respectively;

3) if $\alpha_T > g_m$ gain saturates (infinite current) only at $g = g_m$, no laser effect is possible;

4) if $\alpha_T < g_m$ gain saturates at $g = \alpha_T$, the laser threshold is achieved.

Equation [4.52] is further considered because of its striking difference with every different formulation of the current–gain relationship available in the literature. This is developed in detail in section 4.15, appendix F.

4.7.3. *Measured optical power* P$_{OUT}$

A final step is now required to conclude this chapter: as the ideal current I_{ph} needed to be referred to the measurable current I and to the unavoidable competing current I_{nr}, the emitted optical power P_{TOT} must be brought to refer to the measurable output power P_{OUT}. This is simply done considering that if a fraction α_m of the total losses α_T gives account for the optical power emitted toward the detection system (see section 4.12, appendix C), and that such a system has some coupling and conversion efficiency, globally indicated with η, we can assume that

$$P_{OUT} = \eta P_{TOT} \frac{\alpha_m}{\alpha_T}$$ [4.53]

The global result of the power–current relationship, considering equations [4.42], [4.48] and [4.51], can be expressed in the form

$$P_{OUT} = \eta \frac{h\nu_{peak}}{q} (I - I_{nr}) \frac{\alpha_m}{\alpha_T}$$ [4.54]

This is an expression of the light–current (LI) curve, which is very close to what is usually found in the literature (Coldren et al. 2012; Agrawal and Dutta 1986; Verdeyen 1995), with the relevant difference that it spans the whole injection range, connecting the subthreshold with the laser range through an embedded threshold condition. It is indeed evident that, splitting the current range at $I = I_{th}$ and recalling equations [4.49] and [4.50], one has

$$P_{OUT} = \begin{cases} \eta \dfrac{h\nu_{peak}}{q} \eta_q I \dfrac{\alpha_m}{\alpha_T} & I < I_{th} \\[2ex] \eta \dfrac{h\nu_{peak}}{q} (I - I_{th}) \dfrac{\alpha_m}{\alpha_T} & I > I_{th} \end{cases}$$ [4.55]

Even an analytic expression is available for the whole current range. It may be rather cumbersome but possibly useful for numerical tasks. It gives the inverse function $I(P_{OUT})$, starting from equation [4.54], substituting equation [4.47] for I_{nr} and introducing, for qV, the expression in equation [4.46]. Some handling leads to

$$I = I_0 \frac{P_{OUT}}{P_0} + I_{th0} \left[\frac{1 + \dfrac{\alpha_T}{g_m}}{\sqrt{1 + \dfrac{P_0}{P_{OUT}} \left(1 + \dfrac{\alpha_T}{g_m}\right) - \dfrac{\alpha_T}{g_m}}} \right]^2$$

with $P_0 = \dfrac{h\nu_{peak}}{q} \eta I_0 \dfrac{\alpha_m}{\alpha_T}$ [4.56]

and I_0 and I_{th} are the same parameters defined in equations [4.46] and [4.50], respectively. It is just a matter of direct inspection to verify that equation [4.56] leads to equation [4.55] for very low and very high values of P_{OUT}, while also defining the threshold current in agreement with equation [4.50].

4.8. Resistive effects

4.8.1. *Series resistance* R_S

The most striking difference between the theoretical current–voltage characteristics in Figure 4.15, and experimental measurements, is that the vertical branch corresponding to the laser regime does not exist. It is sufficient to include (Figure 4.16) a series resistance R_S in the schematics of Figure 4.14 and to replot (Figure 4.17) the measured curves, with respect to the reduced voltage $V = V_{ext} - R_S I$ to bring back experiments to the expected behavior.

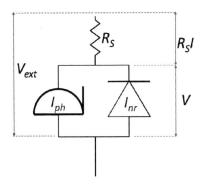

Figure 4.16. *The modified equivalent circuit for a laser diode, including a series resistance R_S*

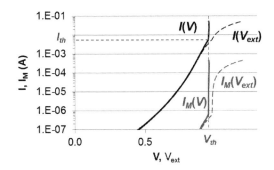

Figure 4.17. *Experimental measurements of the total laser current I and of the photocurrent I_M induced in a photodetector coupled with the laser. The dashed lines have the external voltage V_{ext} as the abscissa. The solid lines replaced that abscissa with the internal voltage $V = V_{ext} - R_S I$. The latter should be compared with Figure 4.15. For a color version of this figure, see www.iste.co.uk/vanzi/reliability.zip*

The value of the series resistance is easily measured from the experimental $I(V_{ext})$ by computing dV_{ext}/dI. Indeed, introducing that resistance, and expressing the voltage V by means of the external voltage, V_{ext}, one has the total current I given by

$$
I(V_{ext}) = \begin{cases} \left(I_{ph0} + I_{nr0}\right)\left(e^{\frac{q(V_{ext}-R_S I)}{kT}} - 1\right) & I < I_{th} \\ \dfrac{V_{ext} - V_{th}}{R_S} & I \ge I_{th} \end{cases}
$$

[4.57]

$$
\frac{dV_{ext}}{dI} = \begin{cases} R_S + \dfrac{kT}{q}\dfrac{1}{I} & I < I_{th} \\ R_S & I \ge I_{th} \end{cases}
$$

[4.58

Figure 4.18 shows the plot of the differential voltage–current characteristics that lead to the determination of R_S and the reconstruction of the internal voltage V from the external voltage V_{ext} allowed by knowing R_S.

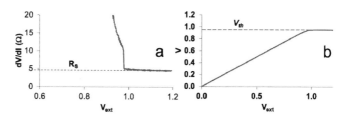

Figure 4.18. (a) The estimate of R_S by means of the differential voltage–current relationship. (b) Internal voltage V and its saturation at V_{th} referred to the external voltage V_{ext}

4.8.2. Side ohmic paths: current confinement

The need for lateral confinement of the injected current leads to several different technological solutions. Each of them implies that some extra current complement the currents I_{ph} and I_{nr} hitherto considered for the ideal device.

The most straightforward way for investigating the general phenomenon and it side implications is to consider a ridge structure, for which even some numerical models can be developed, as proposed in Vanzi (2008), Mura and Vanzi (2010), and Vanzi *et al.* (2011) and summarized in appendix E.

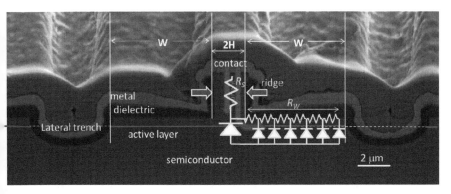

Figure 4.19. *Front view of a InP-based ridge laser diode using the scanning electron microscope, with superimposed line locating the position of the active layer, and part of the distributed electrical equivalent circuit. For a color version of this figure, see www.iste.co.uk/vanzi/reliability.zip*

The kernel is that the lateral wings at both sides of the ridge behave as a transmission line, as shown in Figure 4.19, driving an extra current I_W, whose contribution becomes negligible as far as voltage and current approach the threshold condition.

4.8.3. Leakage paths

Finally, we introduce a possible parallel shunting resistance, which accounts for parasitic paths as surface conduction and is restricted in regular devices to extremely low current ranges. This small last current, that completes the list of all currents, is here indicated as I_{sh}, and is driven by the same internal voltage V across the shunting resistance I_{sh}.

$$I_{sh} = \frac{V}{R_{sh}} = \frac{V_{ext} - R_S I}{R_{sh}} \qquad [4.59]$$

4.8.4. Total current and its components

The (simplified) equivalent circuit is then represented by the schematics in Figure 4.20. Here, the lateral path, responsible for the side current I_W, has been represented by a single resistor-diode couple, which is even an appreciable approximation of the real characteristics for non-degraded devices. In any case, the correct $I_W(W)$ characteristics are those given in appendix E.

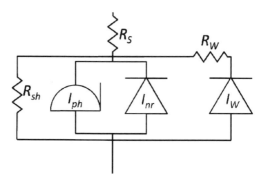

Figure 4.20. *The complete equivalent circuit of a laser diode*

The plot of the various currents and of their sum, the total current I, is shown in Figure 4.21.

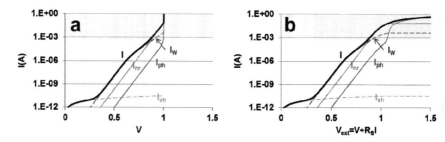

Figure 4.21. *The total current I and its components as functions of (a) the internal voltage V and (b) the external voltage V_{ext}. The yellow regions indicate the current ranges in the mA range and higher. It corresponds to the sole current values that can be appreciated in linear scale, as shown in Figure 4.13. For a color version of this figure, see www.iste.co.uk/vanzi/reliability.zip*

4.9. Non-idealities

Real devices are not ideal, and the previous chapters yet required us to introduce several non-ideal currents, such as the non-radiative, the lateral or the shunting ones in the main body of the model.

Here, the following three cases of non-ideality will be considered:

1) the presence of multiquantum wells (MQWs);

2) the apparent reduction of R_S at increasing current beyond I_{th};

3) the big puzzle of the non-unitary ideality factor.

The choice has also been made not to deal with thermal effects. It is assumed that the available literature on this subject does not need any revision, at least on the basis of the present model.

1) Multiquantum wells

A first modification should deal with MQWs. The model assumes the same approximations that lead, for instance, from equations [2.43] to [2.45] in Coldren et al. (2012). Here, the value of gain calculated for a single quantum well is simply multiplied by the number of wells. In other words, a sort of *effective volume* is introduced, summing all QWs. Within such approximation, currents are the same as in the basic model, apart from the effective volume, that is now NW times the volume of a single well, where NW is the number of wells.

2) Threshold propagation (RS reduction after threshold)

In Figure 4.18a, the actual line representing R_S slightly decreased when voltage increased, as if the series resistance would reduce as injection becomes higher and higher. This can be observed in plots shown in Figure 4.22, where current–voltage values have been collected well beyond the threshold.

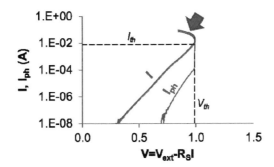

Figure 4.22. *Real current–voltage characteristics for a real device at 980 nm with $I_{th} \approx 10$ mA. This should be compared with Figure 4.17. The measured photocurrent has been scaled up to reconstruct the radiative component I_{ph}. For a color version of this figure, see www.iste.co.uk/vanzi/reliability.zip*

The overlapping part of I and I_{ph}, indicated by the arrow in Figure 4.22, bends to the left for currents increasing beyond the threshold. It looks as if the correction R_S of the abscissa, which converts the external voltage V_{ext} into the internal voltage V would become excessive at higher currents. Once again, it looks as if the series resistance R_S would reduce at high injection.

The observed phenomenon is, on the contrary, explained by that same distributed electrical model that was used for studying the lateral current I_W in Figure 4.19 and in appendix E. Dealing with I_W, indeed, we restricted to the subthreshold range assuming that it makes no sense to build a device that also allows the lateral wings at the sides of the ridge to achieve the laser threshold. Accordingly, the distributed model for lateral current was then made of ideal Shockley diodes.

Once the active area covered by the ridge is at threshold, any further increase of the injected current also leads part of the lateral wings to reach the threshold condition. That part, indicated by the section ΔH in Figure 4.23, will progressively enlarge as injection increases.

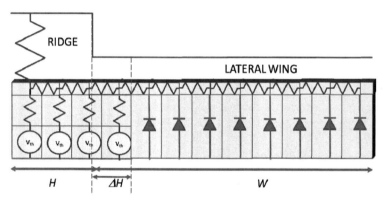

Figure 4.23. *The distributed model for lateral current extended to the above-threshold current range. For a color version of this figure, see www.iste.co.uk/vanzi/reliability.zip*

An analytical model for that situation was not found, but a numerical solution of the circuit shown in Figure 4.23 was obtained by letting ΔH increase from 0 (solid ridge threshold) to ideally W, and calculating for each case the lateral current I_W, the total current I and the applied external voltage V_{ext}.

The result is plotted in Figure 4.24, where the same data of Figure 4.18a were used.

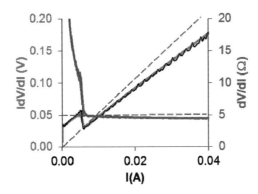

Figure 4.24. *Measured (thin solid lines) and calculated (thick solid lines) differential dV/dI and IdV/dI curves considering lateral threshold propagation. The dashed lines indicate the ideal behavior for both plots. For a color version of this figure, see www.iste.co.uk/vanzi/reliability.zip*

The IdV_{ext}/dI curve, which is also shown in Figure 4.24, is the plot of equation [4.58] multiplied by the current I.

$$I\frac{dV_{ext}}{dI} = \begin{cases} R_S I + \dfrac{kT}{q} & I < I_{th} \\ R_S I & I \geq I_{th} \end{cases} \qquad [4.60]$$

The ideal plot should be made of two segments of parallel lines, vertically separated by a quantity kT/q, with the lower laying on a straight line from the origin. It is an alternative way for observing the apparent reduction of the series resistance R_S.

3) Non-unitary ideality factor

The most intriguing point in the proposed model (Vanzi *et al.* 2015) becomes evident after observing the following features:

– comparing the predicted (Figure 4.15 and Figure 4.21) and the measured I_{ph} (Figure 4.17), one realizes that the former has a smooth transition from the Shockley regime to the voltage clamp, while the latter shows a sharp kink, similar to that displayed by the total current I in both theory and experiments, given by the additive combination of the two different curves I_{nr} and I_{ph};

– the experimental characteristics in the mA subthreshold range of Figure 4.17 are parallel, but their ideality factor n corresponds to some non-unitary value (about

$n = 1.2$, for the edge emitter at 1310 nm in Figure 4.19). Different values of n are met in all devices studied up to now, but always larger than unity. For instance, in a Vertical Cavity Surface Emitting Laser (VCSEL), at 850 nm, the result is $n = 1.82$;

– the threshold voltage V_{th} is a few percent larger than the ideal transparency voltage V_{tr}. In particular, we measured 4% for the edge emitter and 7% for the VCSEL.

The point is that the two last observations are conflicting. Let us consider, for instance, $n = 1.2$ as for the edge emitter at 1310 nm that corresponds to a photon energy of 0.947 eV. Even neglecting any series resistance, the voltage required for reaching transparency should accordingly be 1.2 times of 0.947 V, which is about 1.14 V. But the measured threshold voltage V_{th}, which is, in any case, somewhat larger than the transparency voltage V_{tr}, was 0.984 V. This is perfectly in agreement with the ideal model, with ideality factor at $n = 1$, but not with its measured ideality factor.

The proposed solution is charming, and is proposed in Figure 4.25: the experimental curve for I_{ph} is made by the contribution of (at least) components $I_{phn}(V/n)$ and $I_{phl}(V)$. The first depends on the reduced voltage V/n as expected for MQW diodes, while the second depends on the full voltage drop V. Moreover, the "saturation current" of the first is supposed to be much larger than that of the second, indicating a much more probable spontaneous process. Because of the shift of the threshold voltage for the first component, the second one will be the first to achieve the laser regime. The resulting characteristics then display a subthreshold slope corresponding to a non-unitary ideality factor, a sharper transition than the theoretical I_{ph}, and a threshold voltage corresponding to the ideal case. It should be noted that the bold line in Figure 4.12 is experimental, while the thin lines show plots of the theoretical $I_{ph}(V/n)$ and $I_{ph}(V)$ after suitable choice of their saturation currents.

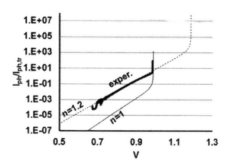

Figure 4.25. *The proposed interpretation of Iph (bold line) for a real 1310 nm emitter based on the hypothesis of dual emission transitions*

The physical interpretation is a real puzzle, whose possible solution has been illustrated extensively in Vanzi et al. (2015). In short (Figure 4.26), it is proposed that, because of the quantum size of the MQW stack, non-local interactions are possible.

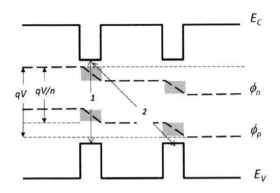

Figure 4.26. *Local and non-local transitions in an ideal double quantum well (Vanzi et al. 2015)*

In particular, transitions of type 1 in Figure 4.26 are assumed to be driven by a reduced separation of the quasi-Fermi levels, but also to be much more probable than type 2, because of perfect overlap of the "local" wavefunctions for electrons and holes. They give rise to a more intense current (radiative and non-radiative) overcoming the much weaker current associated with the transitions between the only partially overlapped wave functions in type 2. But the latter are driven by the full separation qV of the quasi-Fermi levels, and then reach the laser threshold first.

4.10. Appendix A: the anomaly ε

In order to discuss the role of the asymmetry parameter ε, as defined in equation [4.11], we will limit ourselves to the simple case of density of states of an ideal quantum well, which is a step function for both electrons and holes.

$$\begin{cases} g_e = \dfrac{m_e}{\pi\hbar^2} & E_e > E_C, \, 0 \text{ otherwise} \\[4mm] g_h = \dfrac{m_h}{\pi\hbar^2} & E_h < E_V, \, 0 \text{ otherwise} \end{cases} \qquad [4.61]$$

From Table 4.1, we obtain

$$\frac{E_e + E_h}{2kT} = \frac{E_C + E_V}{2kT} + \frac{m_h - m_e}{m_h + m_e} \frac{hv - E_g}{2kT} \qquad [4.62]$$

On the other hand, for an undoped semiconductor the total number of electrons equates to the total number of holes, so that

$$\int_{E_C}^{\infty} n(E_e) dE_e = \int_{-\infty}^{E_V} p(E_h) dE_h \qquad [4.63]$$

Using the definition of carrier densities given in Table 4.1, and equation [4.61] for the densities of states, one obtains

$$\frac{\phi_n + \phi_p}{2kT} = \frac{E_C + E_V}{2kT} + \ln\left(\sqrt{\frac{m_h}{m_e}}\right) = \frac{E_i}{kT} \qquad [4.64]$$

where the independence of the result on the applied bias V allows us to identify the mean value of the quasi-Fermi levels as the Fermi level E_i itself at equilibrium for the undoped material.

The result is given as:

$$\varepsilon = \frac{(E_e + E_h) - (\phi_n + \phi_p)}{2kT} = \frac{m_h - m_e}{m_h + m_e} \frac{hv - E_g}{2kT} - \ln\left(\sqrt{\frac{m_h}{m_e}}\right) \qquad [4.65]$$

It is clear that for ideal symmetric bands (equal effective masses), this value is null.

For asymmetric bands, the two terms on the right-hand side of equation [4.65] partially compensate, depending on the displacement of the peak emission hv with respect to the energy gap E_g.

The maximum absolute value is reached for the non-broadened distribution of states, where that displacement is null, which makes the first of the two terms vanish. In any case, some non-null value of ε must be considered.

The effect of ε has been duly calculated for all the relevant results of the model obtained, using formula [4.75], but it can be summarized as follows:

– photon density:

$$\phi_V - \phi_{0V} = \phi_{0V} \frac{\exp\left(\dfrac{qV}{kT}\right) - 1}{R\left[\exp\left(\dfrac{qV - h\nu}{kT}\right) + 2\exp\left(\dfrac{qV - h\nu}{2kT}\right)\cosh\left(\varepsilon\right) + 1\right] + \left[1 - \exp\left(\dfrac{qV - h\nu}{kT}\right)\right]} \qquad [4.66]$$

– threshold voltage

$$qV_{th} = h\nu + kT \ln\left(\frac{\sqrt{1 + R^2 \sinh^2\left(\varepsilon\right)} + R\cosh\left(\varepsilon\right)}{1 - R}\right)^2 \qquad [4.67]$$

– gain

$$g = g_m \frac{1 - \exp\left(\dfrac{h\nu - qV}{kT}\right)}{\left[\exp\left(\dfrac{h\nu - qV}{kT}\right) + 2\exp\left(\dfrac{h\nu - qV}{2kT}\right)\cosh\left(\varepsilon\right) + 1\right]} \qquad [4.68]$$

The previous considerations have great relevance when one considers not only band asymmetry but also the multiplicity of the valence band, and in particular the existence of the light and heavy hole subbands (the split-off band will here be neglected because of its lower population, assumed to marginally affect the main transition rates). Each subband contributes to optical transitions, giving rise to two specific radiative currents I_{phl} and I_{phh} that add to build up the total I_{ph}.

The light-hole contribution I_{phl} has a lower "saturation current" in the subthreshold range than the heavy-hole term I_{phh}, but displays a lower threshold voltage than the latter.

As an example, a numerical evaluation can be given for the $In_{1-x}Ga_xAs_yP_{1-y}$ lattice-matched to InP to give emission at 1310 nm, as shown in Figures 4.5 and 4.6, applying the parameter evaluation summarized by Agrawal and Dutta (1986, Chapter 3). We get

$$\varepsilon = 0.33\frac{hv - E_g}{2kT} - 0.34 \text{ for transition between conduction band and the light}$$

hole band, and $\varepsilon = 0.79\dfrac{hv - E_g}{2kT} - 1.06$ when, on the contrary, transitions involve

heavy holes.

If we refer to the results with the ideal case $\varepsilon = 0$, for room temperature, $hv = E_g$ = 0.9466 eV, and $R = 0.4$, that would lead to $V_{th} = 0.9905$ V, we obtain the results listed in Table 4.2, where the saturation current, the threshold voltage and the threshold current are referred to the ideal case of symmetric bands.

When we realize that it is the lower threshold voltage that rules over the whole system, this means that threshold is governed by the light-hole transitions. For threshold voltages close to 1 V and threshold currents about 10 mA, the use of the approximation $\varepsilon = 0$ would then give an error of 1 mV on V_{th} and of 0.4 mA on I_{th}. Figure 4.27 draws and compares the three calculated plots of I_{ph}. The symmetric band approximation results are nearly undistinguishable from the light-hole transitions.

	Relative I_{sh0}	Relative V_{th}	Relative I_{th}
$\varepsilon = 0$ (symmetric bands)	1	1	1
$\varepsilon = -0.34$ (light holes)	1.16	1.001	1.04
$\varepsilon = -1.06$ (heavy holes)	1.34	1.012	1.6

Table 4.2. *Relative values of the relevant parameters for the symmetric band model and the asymmetric models involving light and heavy holes*

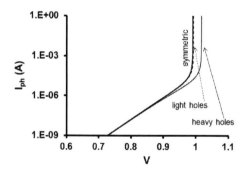

Figure 4.27. *Expected I_{ph} at the laser transition for the symmetric band model, and for the light or heavy hole transitions in a 1310 nm emitter ($In_{1-x}Ga_xAs_yP_{1-y}$ lattice-matched to InP)*

This is the reason for considering the simpler version of the model, corresponding to the non-realistic band symmetry approximation, an excellent tool for practical applications. The only point to be kept in mind is that, for the given case, the subthreshold values of I_{ph} will result some 30% lower in the theoretical model than in experiments.

4.11. Appendix B: optical losses

Losses inside a uniform unlimited domain are only internal losses, simply because an external world does not exist. The situation changes when finite domains, and then boundaries, are taken into account, because photons can *escape*. Problems arise when one tries to include all escape rates into the rate equation.

If we consider, for instance, an edge emitter, the active region is a stripe (Figure 4.28) several hundred micrometers long (L), some nanometer high (d) and a few micrometer large (W).

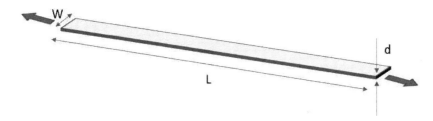

Figure 4.28. *The three dimensions of the active stripe in an edge emitter. Large arrows indicate light output*

Each dimension was granted by a different representation for the respective photon escape.

– Photon escape from the edges of the long side L is the only one that can be modulated by resonances, where length L is the sole dimension that is much larger than the operating wavelengths of the laser diodes. It is so important that a specific term α_m (m stands for "mirror", recalling the partial reflections occurring at the edges of the cavity) has been introduced for that specific escape.

– Lateral escapes from the boundaries of the width W are often simply neglected. They would be worthy of a dedicated coefficient α_L, or it can be included in a new concept of "total internal" losses α_I

$$\alpha_I = \alpha_i + \alpha_L \qquad [4.69]$$

– Vertical escapes from the smallest dimension represent those photons that propagate into the confinement layer. They could be consistently indicated by another coefficient α_V, so that the total losses could be written as

$$\alpha_T = \alpha_i + \alpha_L + \alpha_V + \alpha_m = \alpha_I + \alpha_V + \alpha_m \qquad [4.70]$$

The point is that vertical losses are described everywhere in the literature, by means of the so-called confinement factor Γ, that is a number ranging from 0 to 1, which represents the fraction of the electromagnetic energy of the optical field that is confined inside the active layer, with respect to the vertical dimension.

It enters the statement of the threshold condition for gain that is worth recalling. Laser regime is achieved when the gain function g will equate the total losses – equation [4.36].

This has a physical meaning, is quite logical and is simple.

Nevertheless, it is customary (Coldren *et al.* 2012) to write

$$\Gamma g_{MAX} = \alpha_I + \alpha_m \qquad [4.71]$$

and to attribute to the right-hand side the role of "total losses".

This last formulation will not be adopted in this text, and equation [4.36] will be considered as the sole threshold condition. The simple definition

$$\Gamma = 1 - \frac{\alpha_V}{\alpha_T} \qquad [4.72]$$

leads equations [4.36] and [4.71] to coincide. The respective right-hand sides should be considered two different ways for defining total losses.

Finally, such a definition of Γ holds at and beyond threshold, but remains questionable at lower injection, when spontaneous emission should not be neglected. Also for this reason, the additive form in equation [4.70] will be used, and the confinement factor is never introduced.

4.12. Appendix C: a continuity equation for photons

The definition of gain, equation [4.29], follows the oversimplification of equation [4.27] that neglects both the spontaneous emission and the contribution ϕ_{0V} of the equilibrium photon density.

$$\frac{\partial \phi_V}{\partial x} = \left(g - \alpha_T\right)\phi_V \tag{4.73}$$

whose solution is an exponential decay, until gain is lower than losses,

$$\phi_V(x) = A\exp\left(-\left(\alpha_T - g\right)x\right) \tag{4.74}$$

and a constant density when $g = \alpha_T$

Its results evidence that the characteristic length ℓ (equivalent to the mean path) for photons is considered to be

$$\ell = \frac{1}{\alpha_T - g} \tag{4.75}$$

Equation [4.73] is derived for a uniform and unlimited domain, where the gradient itself should be null.

The point is that in dealing with photons as particles, we surely renounce to wave features, but we should also pretend that the density of such particles obeys the same conservation rules that we know as *continuity*.

In other words, we should expect that our uniform and unlimited domain is a particular case of a more general situation, ruled by a *continuity equation* of the type known for minority carriers in solid-state electronics (Shockley 1950)

$$\frac{\partial \phi}{\partial t} = D\frac{\partial^2 \phi}{\partial x^2} + G - \frac{\phi - \phi_0}{\tau} \tag{4.76}$$

where we recognize the density ϕ, the diffusion term $D\frac{\partial^2 \phi}{\partial x^2}$, that introduces the diffusion coefficient D, a generation term G, and a *loss* term $\frac{\phi - \phi_0}{\tau}$ that is proportional, through the constant $\dfrac{1}{\tau}$ where τ is the lifetime, to the excess density with respect to equilibrium $\phi - \phi_0$. The drift terms that account for motion of charged particles under the action of an electric field have been omitted.

Several important differences should be taken into account:

– Because of the k-selection rule, the balance must be detailed for each photon frequency. On the contrary, minority carriers are considered in their totality.

– Photon generation may occur both independently (spontaneous emission, with rate R_{sp}) of the existing density, and then identifying the term G, and proportionally (stimulated emission) to the density ϕ. This last phenomenon would introduce a positive term (not existing for minority carriers) with the same characteristics of the last term.

– Photon absorption, which belongs to the destruction phenomena, and then to the last term, is traditionally considered the competitor of the stimulated emission and their joint effect enter the definition of optical gain. For this reason, the last term must, in any case, be split into an absorption and a loss term.

– The total effect of the last considerations is that the lifetime is conveniently replaced by $c(\alpha - g)$, where c is the light speed inside the optical material, α is the loss coefficient that accounts for all loss and escape phenomena different from photon absorption, and g is the optical gain that merges creation and destruction of photons by means of stimulated transitions. It is important to recall that the term $(\alpha - g)$ has the dimensions of the reciprocal of a length, and that such a length has the meaning of mean free path for photons.

– After rearranging the last two addenda in terms of gain and losses, their coefficients also lead to the diffusion coefficient. It is defined, as usual, as the product of the particle speed (in our case the light speed in the active material c) and its mean free path ℓ. Because of the previous point, we must assume

$$D = \frac{c}{\alpha - g}$$
[4.77]

The continuity equation for photons is then expected to be:

$$\frac{\partial \phi_v}{\partial t} = \frac{c}{\alpha - g} \frac{\partial^2 \phi_v}{\partial x^2} + R_{sp} + cg\phi_v - c\alpha\left(\phi_v - \phi_{v0}\right)$$
[4.78]

In contrast to equation [4.27], it is now possible to consider the steady state without also forcing the gradient to be null

$$0 = \frac{\partial^2 \phi_V}{\partial x^2} + \left[\frac{R_{sp}}{c} + g\phi_{V0}\right](\alpha - g) - (\alpha - g)^2 (\phi_V - \phi_{V0}) \tag{4.79}$$

The term in square brackets, which can be verified to vanish when the system approaches equilibrium, can be alternatively rewritten by means of our notations as

$$\left[\frac{R_{sp}}{c} + g\phi_{V0}\right] = \phi_{0V} g_m \frac{\exp\left(\dfrac{qV}{kT}\right) - 1}{\left[1 + \exp\left(\dfrac{qV - hv}{2kT}\right)\right]^2} = \phi_{0V} g_m \left\{\frac{e^{\frac{hv}{kT}} - 1}{4}\left(1 + \frac{g}{g_m}\right)^2 + \frac{g}{g_m}\right\} \tag{4.80}$$

What is interesting now is that equation [4.79] has the general solution

$$\phi_V - \phi_{V0} = \frac{\left[\dfrac{R_{sp}}{c} + g\phi_{V0}\right]}{\alpha - g} + K_1 \exp\big((\alpha - g)x\big) + K_2 \exp\big(-(\alpha - g)x\big) \tag{4.81}$$

and K_1 and K_2 are constants to be defined according to the boundary conditions.

– For the **infinite domain**, both constants K_1 and K_2 must be null in order to keep the solution finite everywhere. The result is a constant value independent on the x coordinate.

$$\phi_V - \phi_{V0} = \frac{\left[\dfrac{R_{sp}}{c} + g\phi_{V0}\right]}{\alpha - g} \tag{4.82}$$

This is easily shown to be exactly the same result as in equation [4.28], once the coefficient α is identified with the total losses α_T, that, for the infinite domain, are only internal losses.

– For a **semi-infinite domain**, defined for instance for $x \geq 0$, the solution must force the sole K_1 to vanish

$$\phi_V - \phi_{V0} = \frac{\left[\dfrac{R_{sp}}{c} + g\phi_{V0}\right]}{\alpha - g} + K_2 \exp\big(-(\alpha - g)x\big) \tag{4.83}$$

Because of the boundary, we must consider that photons can now leave the domain at the facet. We can assume, as in Grove (1967) dealing with *surface recombination* of minority carriers in semiconductor, that the photon flux leaving the edge $x = 0$ is *proportional* to the photon density at that same edge. Moreover, the flux itself, being a diffusion phenomenon, is in turn proportional to the density gradient, measured at that same point. The two statements can be summarized in the boundary condition

$$\frac{\partial \left(\phi_V - \phi_{V0} \right)}{\partial x}\bigg|_{x=0} = \alpha_m \left(\phi_V - \phi_{V0} \right)\big|_{x=0} \qquad [4.84]$$

where α_m is the proportionality coefficient that has the dimension ℓ^{-1} of a loss coefficient.

The final result for the photon density is intriguing

$$\phi_V(x) - \phi_{V0} = \frac{\left[\dfrac{R_{sp}}{c} + g\phi_{V0} \right]}{\alpha - g} \left[1 - \frac{\alpha_m}{\alpha + \alpha_m - g} \exp\left[-(\alpha - g)x \right] \right] \qquad [4.85]$$

This is a function that at large distance from the boundary $x = 0$ behaves as the solution for the unlimited domain, equation [4.82], and has an exponential decrease approaching the edge of the domain.

But the condition for an unlimited photon emission shifts forwards: as far as g approaches α, one has

$$\phi_V(x) - \phi_{V0} = \frac{\left[\dfrac{R_{sp}}{c} + g\phi_{V0} \right]}{\alpha - g} \left[1 - \frac{\alpha_m}{\alpha + \alpha_m - g} \exp\left[-(\alpha - g)x \right] \right] \xrightarrow{g \to \alpha} \frac{\left[\dfrac{R_{sp}}{c} + g\phi_{V0} \right]}{\alpha + \alpha_m - g} \qquad [4.86]$$

In the same way, the outgoing flux is always proportional to

$$\alpha_m \left(\phi_V(0) - \phi_{V0} \right) = \alpha_m \frac{\left[\dfrac{R_{sp}}{c} + g\phi_{V0} \right]}{\alpha + \alpha_m - g} \qquad [4.87]$$

In other words, the role played by the sole α in the unlimited domain is now transferred to the sum $\alpha + \alpha_m$, that assumes the meaning of *total losses* and is worthy of a specific coefficient, α_T :

$$\alpha_T = \alpha + \alpha_m \tag{4.88}$$

We can now recognize the twofold expression of optical loss: a uniformly distributed phenomenon, that we can call *internal losses*, identified by α, and the *mirror losses*, identified as α_m, that describe photon escape from the "open" boundaries of the domain.

This is the seed for the construction of the total loss coefficient as given in equation [4.70], and is the way to consider escape losses even from reflection-less interfaces.

4.13. Appendix D: the integral of the spectral function

The integral in equations [4.40] and [4.41] can be restricted to the sole photon density, assuming also for the coefficient τ_C (and then for the loss term defined in equation [4.26]) to be a slow function of frequency.

$$\int_0^\infty \frac{\phi_\nu - \phi_{0\nu}}{\tau_C} d\nu \approx \frac{1}{\tau_C} \int_0^\infty [\phi_\nu - \phi_{0\nu}] d\nu \tag{4.89}$$

We shall first assume α_T constant and g_m an ideal step function, centered at $h\nu = E_g$.

Let us start with the range of voltages for which the approximation in equation [4.30] holds

$$\int_0^\infty [\phi_\nu - \phi_{0\nu}] d\nu = \frac{\exp\left(\dfrac{qV}{kT}\right) - 1}{\dfrac{\alpha_T}{g_m} + 1} \int_{h\nu = E_g}^\infty \phi_{0\nu} d\nu \tag{4.90}$$

Because the exponential in the Planck's function, equation [4.16], by far dominates over the unit, even at the lower limit $h_v = E_g$, one easily finds that the integral in equation [4.90] is simply kT/q times the value of the integrand ϕ_{0v} calculated at $hv = E_g$. In other words, for low voltages

$$\int_0^\infty \left[\phi_v - \phi_{0v}\right] dv = \frac{kT}{q}\left[\phi_v - \phi_{0v}\right]_{peak} \qquad [4.91]$$

On the other hand, keeping the same ideal step function for g_m, the integral of the excess photon density $\phi_v - \phi_{0v}$ over the range $hv \geq E_g$, after the substitution

$X = \exp\left(\dfrac{qV - hv}{2kT}\right)$, becomes

$$\int_0^\infty \left[\phi_v - \phi_{0v}\right] dv = \frac{A}{B}\frac{2kT}{h} \int_0^{\exp\left(\frac{qV - E_g}{2kT}\right)} \frac{X}{R(1+X)^2 + 1 - X^2} dX \qquad [4.92]$$

with R now a constant value across the integration range. Without going to the explicit logarithmic form of the solution, that is easily demonstrated to converge to equation [4.91] for $qV < E_g$, it results to diverge at that same value $X = \dfrac{1+R}{1-R}$ at which the integrand diverges.

In other words, for α_T constant and g_m an ideal step function, centered at $hv = E_g$, the integral behaves as the integrand calculated at its peak. Figure 4.29 shows that the cumulative contribution of four densities calculated for this case simply replicates the behavior of the peak photon density.

Some difference will be displayed in the transition region that in Figure 4.4 connects the two straight branches close to $V = 1$ V, when the actual spectral behavior of α_T and g_m is not that simple. Figure 4.30 illustrates what changes if we assume that the photon density has a different spectral distribution than in the previous case.

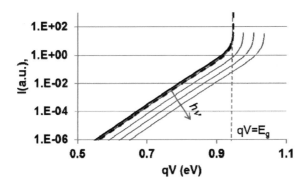

Figure 4.29. *Photon density at four different photon energies (thin lines) increasing from $h\nu = E_g$ (leftmost thin line) by steps of 0.03 eV, in accordance with the step model for g_m and $\alpha_T/g_m = 0.5$ for $h\nu > E_g$. The thick solid line is the sum of the four thin lines, while the thick dashed line is the same sum suitably scaled to show its perfect coincidence with the sole peak photon density at $h\nu = E_g$. For a color version of this figure, see www.iste.co.uk/vanzi/reliability.zip*

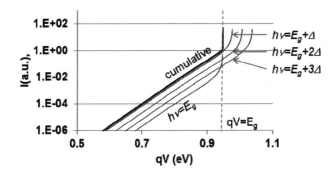

Figure 4.30. *The same as Figure 4.29, with the sole density at $h\nu = E_g$ scaled down by a factor of 100. The cumulative curve now displays the same threshold as before, and preserves the same slope of each spectral density in the subthreshold range. The sole difference is the more sharp transition at threshold. For a color version of this figure, see www.iste.co.uk/vanzi/reliability.zip*

The case proposed in Figure 4.29 is completely arbitrary with respect to the relative intensity of the single spectral lines, but points out as a non-constant ratio α_T/g_m in the allowed photon spectral range will be likely to affect the sole transition region of the cumulative curves in the proximity of the threshold condition.

4.14. Appendix E: the lateral current I_W

In a ridge structure as in Figure 4.19, the epitaxial stack is the same across the whole device, and the double heterostructure (or possibly the set of MQW) is located very close to the top surface (about 1 μm in this case). The lateral trenches and the dielectric prevent the most part of the semiconductor from coming into contact with the upper metal layer. The central ridge is the sole part of the semiconductor where the dielectric is not present, and then acts as a funnel for the flux of charges. In a well-designed device, such structure leads most of the current to cross the junction just under the ridge, allowing a small part of residual current to flow laterally at some extent, before crossing the junction. Such a distribution is indicated as current confinement, and is an effective way of getting low threshold currents and, even more important, lateral photon confinement.

We can represent the electrical equivalent structure by means of a distributed model, and study the behavior of the whole system. It should be clear that the same voltage V that drives the central diode, corresponding to the active region under the ridge, also drives the lateral wings. This voltage is related to the external, directly measurable voltage by the simple relation $V = V_{ext} - R_S I$.

4.14.1. Subthreshold range and current confinement

The first range of interest is $V < V_{th}$. Under the electrical point of view, we have no voltage clamp anywhere, and all diodes behave as Shockley diodes, as indicated in equation [4.57] for the $I < I_{th}$ range.

The DC current flowing in the lateral wings can be calculated assuming that a wing extends from $x = 0$, just at the side of the ridge up to $x = W$, the boundary of the trench shown in Figure 4.19.

Let V be the voltage across the first diode of the wing in $x = 0$ and I_{W0} the current entering the wing at that point, which is the total current consumed by the wing. We assume dx as the lateral extension of each element, R_L for the total resistance of the lateral path extending from 0 to W and I_{WS} for the saturation current of the collective lateral diode in case of null lateral resistances. Assuming each diode driven by Shockley current, and neglecting the unity in comparison with the exponential, we can write for the lateral currents and voltages

$$\begin{cases} \dfrac{d}{dx} V(x) = -\dfrac{R_W}{W} I_W(x) \\ \dfrac{d}{dx} I_W(x) = -\dfrac{I_{WS}}{W} \exp\left(\dfrac{qV(x)}{kT} \right) \end{cases} \qquad [4.93]$$

Differentiating the second equation in [4.93], and substituting dV/dx with the first equation, one has

$$\frac{d^2}{dx^2} I_W(x) = -\frac{q}{kT} W I_W(x) \frac{d}{dx} I_W(x)$$
[4.94]

whose first integration is straightforward

$$\frac{d}{dx} I_W(x) = -\frac{q}{2kT} \frac{R_W}{W} \left(I_W^2(x) + \gamma^2 \right)$$
[4.95]

where γ^2 is an integration constant with the dimensions of a current.

The solution, as shown in Vanzi (2008) and Mura and Vanzi (2010), must lead the lateral current I_W to vanish at the end of the wing, in $x = W$, and then

$$I_W(x) = \gamma \tan\left[\gamma \frac{q}{2kT} R_W \left(1 - \frac{x}{W} \right) \right]$$
[4.96]

Now, γ is related to the boundary condition in $x = 0$, where the current is I_{W0}

$$I_{W0} = \gamma \tan\left[\gamma \frac{q}{2kT} R_W \right] \approx \gamma^2 \frac{q}{2kT} R_W$$
[4.97]

The approximation of the tangent with its argument has been introduced in order to have an insight of the result. We can indeed substitute this value of γ^2 in equation [4.95], and also replace dI_W/dx with the second of equation [4.93] to get the current–voltage relationship in $x = 0$. The ultimate result is

$$I_{W0} = \frac{kT}{qR_W} \left(\sqrt{1 + 2\frac{q}{kT} R_W I_{WS} \exp\left(\frac{qV}{kT}\right)} - 1 \right)$$
[4.98]

It is interesting to observe the opposite extremal behaviors for low or high V.

$$I_{W0} = \begin{cases} I_{WS} \exp\left(\dfrac{qV}{kT}\right) & low\ V \\[4mm] \left[\sqrt{2\dfrac{kT}{qR_W} I_{WS}} \right] \exp\left(\dfrac{qV}{2kT}\right) & high\ V \end{cases}$$
[4.99]

First of all, we see that for very low currents (when the ohmic drops along the lateral, resistors are expected to be negligible), the circuit behaves as the parallel of all lateral diodes. In particular, being that the side areas are much larger than the area limited under the ridge, the lateral current I_{W0} will be much larger than the current flowing under the ridge. But as V increases, the lateral resistive path will limit the lateral current, which asymptotically approaches the slope of an ideal diode with a doubled ideality factor.

The physical interpretation of the low voltage range of I_W allows us to refer it to the current flowing in the active region, which we will shortly indicate as I_A, a current that represents the global Shockley current flowing across the active region below threshold, and then can be identified with the corresponding first line of equation [4.57], expressed in terms of the internal voltage $V = V_{ext} - R_S I$.

$$I_A(V) = \left(I_{ph0} + I_{nr0} \right) \left(e^{\frac{qV}{kT}} - 1 \right)$$

[4.100

Referring to Figure 4.19, we can attribute to the lateral current a saturation proportional to its relative lateral extension, with respect to the ridge:

$$I_{WS} = \left(I_{ph0} + I_{nr0} \right) \frac{W}{H}$$

[4.101

Figure 4.31 draws the relative contribution of the lateral current I_W and I_A to the total current I in the subthreshold range.

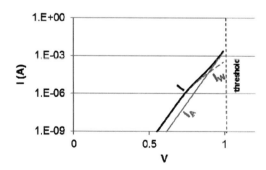

Figure 4.31. *Subthreshold contributions of the currents I_A, flowing across the active region, and I_W, flowing along the lateral wings of a ridge structure, to the total current I. For a color version of this figure, see www.iste.co.uk/vanzi/reliability.zip*

In a well-designed device, although the lateral current dominates at low voltages, it becomes negligible, with respect to the current I_A in the active region, before the internal voltage V approaches the threshold value V_{th}.

For instance, in a device with a threshold current of some few mA, the lateral current falls I_W below I_A in the sub-mA range. This is the reason for which, plotting currents in linear scale as for the LI curve in Figure 4.13, all effects of lateral currents collapse in a negligible region close to the origin.

Nevertheless, the role of the lateral wings, and of their resistive current paths, is crucial for total current to focus into the active region as the injection level increases. It is the current confinement required for also keeping the light emission confined in the laser regime.

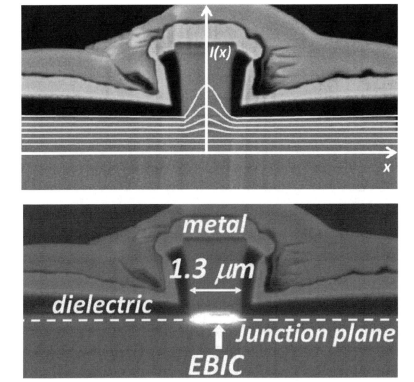

Figure 4.32. *Upper image: qualitative representation of current confinement at increasing injection. Lower image: experimental evidence of current confinement at the SEM. For a color version of this figure, see www.iste.co.uk/vanzi/reliability.zip*

This current confinement is experimentally observable using scanning electron microscopy (Vanzi *et al.* 2016) by means of the electron beam induced current (EBIC) technique that has the property of mapping a charge distribution, demonstrated to be proportional to the current distribution of a forward biased junction.

4.15. Appendix F: the gain–current relationship and its comparison with the literature

4.15.1. *Gain equations in the literature*

Several expressions have been proposed for single-mode gain, starting from the theoretical form that links that quantity to both frequency and quasi-Fermi levels (Lasher and Stern 1964; Verdeyen 1995), which means dealing with both spectral and modal gain.

Referring to Verdeyen (1995), the expression for gain is

$$g = A \frac{\lambda_0^2}{8\pi n_r^2} \left[f_e \left(1 - f_h \right) \right] N_V^2 \left[1 - \exp\left(\frac{h\nu - \left(\phi_n - \phi_p \right)}{kT} \right) \right] \qquad [4.102]$$

Here, the notations of Vanzi (2008), Mura and Vanzi (2010), and Vanzi *et al.* (2011) are used, for the sake of the next comparison with the new formulas, but the term-to-term correspondence with the quoted reference holds. In particular, g is the gain function, A is the Einstein's coefficient for spontaneous emission, λ_0 is the photon wavelength in vacuum, n_r is the refractive index of the active region, f_e and f are the Fermi distribution for, respectively, electrons and holes, N_V^2 is the joint density of states, $h\nu$ is the photon energy, kT is the thermal energy, and ϕ_n and ϕ_p are the quasi-Fermi levels for electrons and holes, respectively.

Equation [4.102] predicts both a lower and upper limit for g. In a specific note in Verdeyen (1995), those limits are shown to have the same absolute value and opposite sign, stating that, on the opposite injection limits, "what was absorption becomes gain". This is not true. On the one hand, the low-injection approximation (unpumped material), which assumes a completely filled state in the valence band and a completely empty state in the conduction band, is acceptable, together with the conclusion that gain in the unpumped material is negative, and coincides with its natural absorption. But the upper limit requires a completely filled state in the conduction band and a completely empty state in the valence band, a physically impossible situation. Moreover, the "freezing" of the quasi-Fermi levels in neglected. This is the only equation that indicates some upper and lower bounds for gain.

The strictly single-mode gain appears in empirical formulas, summarized in Coldren *et al.* (2012). Two main models are available: linear and logarithmic.

1) *The linear model*

Two popular versions do exist; one refers to charge density n and its value n_{tr} at transparency

$$g = a(n - n_{tr})$$ [4.103]

and the other refers to current density J and the corresponding J_{tr} at transparency.

$$g = b(J - J_{tr})$$ [4.104]

The two hardly match because they would imply a current linearly proportional to the charge density. This proportionality is typical of defect-driven recombination, and then to non-radiative mechanisms. Band-to-band transitions, on the contrary, are proportional to the product of the electron–hole densities, that is to n^2.

2) *The logarithmic model*

For the logarithmic model, two versions do exist: charge and current densities

$$g = g_{0n} \ln\left(\frac{n}{n_{tr}}\right)$$ [4.105]

$$g = g_{0J} \ln\left(\frac{J}{J_{tr}}\right)$$ [4.106]

Here, g_{0n} and g_{0J} are introduced as "fitting parameters", without any deeper meaning.

Furthermore, in order to avoid divergence for low vanishing low values of n or J, a version appears that introduces some phenomenological terms n_S and J_S:

$$g = g_{0n} \ln\left(\frac{n + n_S}{n_{tr} + n_S}\right)$$ [4.107]

$$g = g_{0J} \ln\left(\frac{J+J_S}{J_{tr}+J_S}\right)$$ [4.108]

although it is noted that, for best curve fitting close to transparency, the first version is preferable.

Several problems arise when one tries to apply the proposed formulas, and even more when one tries to harmonize them.

Equations [4.103], [4.105] and [4.107], based on the carrier density n, may include gain saturation, provided a similar saturation for n is forced from outside, a way of shifting the problem. The lower limit is predicted by the linear model (equation [4.103]), but not by the pure logarithmic model (equation [4.105]), and the empirical introduction of the term n_S in equation [4.107] has no physical basis.

Even worse, the current-based equations [4.104], [4.106] and [4.108] share the same problems for the lower limit, but in no way can include gain saturation because current is known to have no upper limit.

Some simplifications come from Coldren et al. (2012) that recognize the linear model as an approximated version of a better logarithmic form. Indeed, the first-order expansion of the pure logarithmic model leads to the linear one, also eliminating the mismatch between the charge-based and the current-based expression

$$g = g_{0n} \ln\left(\frac{n}{n_{tr}}\right) \approx \frac{g_{0n}}{n_{tr}}\left(n-n_{tr}\right)$$ [4.109]

$$g = g_{0J} \ln\left(\frac{J}{J_{tr}}\right) \approx \frac{g_{0J}}{J_{tr}}\left(J-J_{tr}\right)$$ [4.110]

Moreover, if one assumes the bi-molecular (band-to-band) recombination as dominating, the relationship between the "fitting coefficients" g_{0n} and g_{0J} is defined

$$\frac{g_{0J}}{g_{0n}} = 2$$ [4.111]

One can conclude that the pure logarithmic form in equation [4.106] is the best reference term for comparison with the new formulas.

Let us then rewrite it as a function of the total current, also simplifying the notation of the fitting coefficient as g_0, and introduce the total current I instead of its density J.

$$g = g_0 \ln\left(\frac{I}{I_{tr}}\right) \qquad [4.112]$$

This also implies the definition of the threshold current I_{th} as the value of I when gain g balances the total optical losses α_T exactly.

$$I_{th} = I_{tr} \exp\left(\frac{\alpha_T}{g_0}\right) \qquad [4.113]$$

4.15.2. Comparison

For the sake of comparison of the present model with the literature results summarized above, the following alternative form for the gain formulas in the main text, obtained by equations [4.33] and [4.47] and the definition of I_{th0} in equation [4.50], will be useful:

$$g = g_m\left(\frac{\sqrt{I_{nr}} - \sqrt{I_{th0}}}{\sqrt{I_{nr}} + \sqrt{I_{th0}}}\right) \qquad [4.114]$$

This formula links gain g to the sole non-radiative current I_{nr} (and its value I_{th0} at transparency), but hides the saturation limit. This is implicitly recovered when one realizes that the first term on the right-hand side of equation [4.52] is exactly I_{nr}; in agreement with its limit, equation [4.50] defines the threshold current I_{th}

$$I_{nr} = I_{th0}\left[\frac{1+\dfrac{g}{g_m}}{1-\dfrac{g}{g_m}}\right]^2 \xrightarrow{g \to \alpha_T} I_{th} \qquad [4.115]$$

Equation [4.114] has a practical importance because of the mentioned dominance of I_{nr} in the whole subthreshold range that includes the transparency value I_{th0}.

This dominance may lead to identify I_{nr}' with the total current I when empirical formulas are looked for. In other words, the form

$$g \approx g_m \left(\frac{\sqrt{I} - \sqrt{I_{tr}}}{\sqrt{I} + \sqrt{I_{tr}}} \right)$$

[4.116]

where I_{tr} is the value of the total current I at transparency, may be considered a practical approximation of equation [4.114] when the accuracy in current measurements is not better than approximately 1%. In this way, the given result for g can be compared with the various expressions, available in the literature, that refer to the total current I.

For the sake of completeness, we must recall that gain may be expressed not only as a function of the current I, but also as a function of the electron density n that has always been assumed equal to the hole density p. Assuming

$$I_{nr} \propto pn = n^2$$

[4.117]

that means the dominance of bi-molecular recombination also for the main component of the subthreshold current, this gives another form of equation [4.116]

$$g = g_m \left(\frac{n - n_{tr}}{n + n_{tr}} \right)$$

[4.118]

There is an evident and striking difference between equations [4.112] and [4.116] and between equations [4.113] and [4.50] and the wide experimental confirmation of equation [4.113] seems to ultimately support the validity of the formers with respect to the latters. On the other hand, the new formulas account for gain saturation, while equation [4.112] seems to allow g to increase with I without any limit. It seems hard to take advantage of the benefits of the two versions.

The surprising result comes from the power expansion of the competing formulas. Let us start with equations [4.112] and [4.116]: they agree mathematically only for the common prediction that gain vanishes at transparency $I = I_{tr}$.

Anyway, close to transparency, we have for equation [4.112]:

$$g = g_o \ln\left(\frac{I}{I_{tr}}\right) = g_o \left[\begin{array}{c} \left(\dfrac{I}{I_{tr}}-1\right) - \dfrac{1}{2}\left(\dfrac{I}{I_{tr}}-1\right)^2 + \dfrac{1}{3}\left(\dfrac{I}{I_{tr}}-1\right)^3 \\ -\dfrac{1}{4}\left(\dfrac{I}{I_{tr}}-1\right)^4 + O\left(\dfrac{I}{I_{tr}}-1\right)^5 \end{array}\right] \qquad [4.119]$$

while for equation [4.116]:

$$g = g_m \left[\dfrac{\sqrt{\dfrac{I}{I_{tr}}}-1}{\sqrt{\dfrac{I}{I_{tr}}}+1}\right] = \dfrac{g_m}{4} \left[\begin{array}{c} \left(\dfrac{I}{I_{tr}}-1\right) - \dfrac{1}{2}\left(\dfrac{I}{I_{tr}}-1\right)^2 + \dfrac{5}{16}\left(\dfrac{I}{I_{tr}}-1\right)^3 \\ -\dfrac{7}{32}\left(\dfrac{I}{I_{tr}}-1\right)^4 + O\left(\dfrac{I}{I_{tr}}-1\right)^5 \end{array}\right] \qquad [4.120]$$

It follows that if we identify the phenomenological fitting parameter g_0 with

$$g_0 = \frac{g_m}{4} \qquad [4.121]$$

any experiment that can confirm equation [4.112] in its range of validity, near transparency, will also confirm equation [4.116] that, in turn, is likely not to differ numerically from its exact formulation, given by equation [4.114].

Being the threshold current I_{th} close to I_{tr}, this conclusion also applies to experiments involving it.

This last particular case has been investigated in a previously (Vanzi *et al.* 2013) where optical losses were modified on a 1310 nm edge emitter by means of a focused ion beam and I_{th} was monitored. The results aligned the experimental measurements, on a suitable plot, all within the range where the two competing formulas for $g(I)$ are completely undistinguishable.

A final comment must be given to equation [4.121]: it assigns a physical meaning to the phenomenological "fitting parameter" g_0 (Coldren *et al.* 2012). The factor 1/4 is not accidental: it corresponds to the transparency value of the absorption term in equation [4.27], so that g_0 assumes the role of the absorption coefficient measured at transparency, that is, when absorption balances stimulated emission exactly.

4.16. References

Adams, M.J. (1973). Rate equations and transient phenomena in semiconductor lasers. *Opto Electronics Review*, 5, 201.

Agrawal, G.P., and Dutta, N.K. (1986). *Long-Wavelength Semiconductor Lasers*. Van Nostrand Reinhold, New York.

Arnold, G., Russer, P., and Petermann, K. (1982) Modulation of laser diodes. In *Semiconductor Devices for Optical Communication*, Kressel, H. (ed.). Springer Verlag, Berlin.

Bloch, F. (1928). Über die Quantenmechanik der Elektronen in Kristallgittern. *Zeitschrift für Physik*, 52, 555.

Bores, P.M., Danielsen, M., and Vlaardingerbroek, M.T. (1975). Dynamic behaviour of semiconductor lasers. *Electronics Letters*, 11, 206.

Casey, H.C., and Panish, M.B. (1978). *Heterostructure Lasers, Part A*. Academic Press, New York.

Coldren, L.A., Corzine, S.W., and Masanovic, M.L. (2012). *Diode Lasers and Photonic Integrated Circuits*, 2nd edition. John Wiley & Sons, Hoboken.

Dirac, P.A.M. (1926). On the theory of quantum mechanics. *Proceedings of the Royal Society A*, A112, 661.

Einstein, A. (1916). Strahlungs-Emission und - Absorption nach der Quantentheorie [Emission and Absorption of Radiation in Quantum Theory]. DPG-Physik, 18, 318–323.

Einstein, A. (1917). Zur Quantentheorie der Strahlung [On the Quantum Theory of Radiation], *Zeitschrift für Physik*, 18, 121–128.

Fermi, E. (1926). Zur Quantelung des idealen einatomigen Gases. *Zeitschrift für Physik*, 36, 902.

Grove, A.S. (1967). *Physics and Technology of Semiconductor Devices*. Wiley, New York.

Henry, C.H. (1985). Spectral properties of semiconductor lasers. In *Semiconductors and Semimetals*, Tsang, W.T. (ed). Academic Press, New York.

Ikegami, T., and Suematsu, Y. (1967). Resonance-like characteristics of the direct modulation of a junction laser. *IEEE Lett.*, 55, 122.

Ikegami, T., and Suematsu, Y. (1968). Direct modulation semiconductor junction lasers. *Electronics and Communications in Japan*, B51, 51.

Kressel, H., and Butler, J.K. (1977). *Semiconductor Lasers and Heterojunction LEDs*. Academic Press, New York.

Lamb, W.E. (1964). Theory of an optical maser. *Physical Review*, 134, A1429.

Lasher, G.J. (1964). Analysis of a proposed bistable injection laser. *Solid State Electronics*, 7, 707.

Lasher, G., and Stern, F. (1964). Spontaneous and stimulated recombination in semiconductors. *Physical Review*, 133(2), A553–A563.

Lau, K., and Yariv, A. (1985). High-frequency current modulation of semiconductor lasers. In *Semiconductors and Semimetals*, Tsang, W.T. (ed.). Academic Press, New York.

Maiman, T.H. (1960). Stimulated optical radiation in ruby. *Nature*, 187(4736), 493–494.

Mura, G., and Vanzi, M. (2010). The interpretation of the DC characteristics of LED and laser diodes to address their failure analysis. *Microelectronics Reliability*, 50(4), 471–478.

Mura, G., Vanzi, M., Marcello, G., and Cao, R. (2013). The role of the optical trans-characteristics in laser diode analysis. *Microelectronics Reliability*, 53(9–11), 1538–1542.

Paoli, T.L., and Barnes, P.A. (1976). Saturation of the junction voltage in stripe-geometry (AlGa)As double-heterostructure junction lasers. *Applied Physics Letters*, 28(12), 714–716.

Paoli, T.L., and Ripper, J.E. (1970). Direct modulation of semiconductor lasers. *Proceedings of the IEEE*, 58, 1457.

Pauli, W. (1925). Über den Zusammenhang des Abschlusses der Elektronengruppen im Atom mit der Komplexstruktur der Spektren. *Zeitschrift für Physik*, 31, 765.

Salathé, R., Voumard, C., and Weber, H. (1974). Rate equation approach for diode lasers. *Opto-Electronics Review*, 6, 451.

Schawlow, A.L., and Townes, C.H. (1958). Infrared and optical masers. *Physical Review*, 112(6), 1940–1949.

Shockley, W. (1950). *Electrons and Holes in Semiconductors*. Van Nostrand, Princeton.

Statz, H., and De Mars, G. (1960). Transients and oscillation pulses in masers. In *Quantum Electronics: A Symposium*, Townes, C.H. (ed.). Columbia University Press, New York.

Statz, H., Tang, C.L., and Lavine J.M. (1964). Spectral output of semiconductor lasers. *Journal of Applied Physics*, 35, 2581.

Thompson, J.H.B. (1980). *Physics of Semiconductor Laser Devices*. John Wiley & Sons, Hoboken.

Townes, C.H., Gordon, J.P., and Zeiger, H.J. (1954). Molecular microwave oscillator and new hyperfine structure in the microwave spectrum of NH3. *Physical Review*, 95(1), 282–284.

Vanzi, M. (2008). A model for the DC characteristics of a laser diode. *ICECS2008*, La Valletta, Malta, pp. 874–877.

Vanzi, M., Mura, G., Marcello, G., Bechou, L., Yannick, D., Le Gales, G., and Joly, S. (2016). Extended modal gain measurement in DFB laser diodes. *IEEE Photonics Technology Letters*. DOI: 10.1109/LPT.2016.2633440

Vanzi, M., Mura, G., Marcello, G., and Martines, G. (2015). Clamp voltage and ideality factor in laser diodes. *Microelectronics Reliability*, 55(9–10), 1736–1740.

Vanzi, M., Mura, G., and Martines, G. (2011). DC parameters for laser diodes from experimental curves. *Microelectronics Reliability*, 51(9–11), 1752–1756.

Vanzi, M., Mura, G., Martines, G., and Tomasi, T. (2013). Optical losses in single-mode laser diodes. *Microelectronics Reliability*, 53(9–11), 1529–1533.

Vanzi, M., Xiao, K., Marcello, G., and Mura, G. (2016). Side mode excitation in single-mode laser diodes. *IEEE Transactions on Device and Materials Reliability*, 16(2). DOI 10.1109/TDMR.2016.2539242

Yariv, A. (1995). *Optical Electronics*, 4th edition. Oxford University Press, Oxford.

Laser Diode DC Measurement Protocols

5.1. The standard *LIV* curve: voltage or current driving

The starting point is the acquisition of DC data, as usual for drawing the so-called *LIV* plot, as shown in Figure 5.1, where current *I*, optical power P_{OUT} and the externally applied voltage V_{ext} are drawn together.

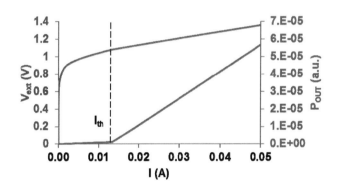

Figure 5.1. *The standard* LIV *plot, relating power, current and voltage in a laser diode. For a color version of this figure, see www.iste.co.uk/vanzi/reliability.zip*

In the standard procedure, the data for drawing Figure 5.1 are obtained by driving the laser diode with a linear current ramp, ranging from 0 to the nominal I_{th}, and measuring $V_{ext}(I)$ and $P_{OUT}(I)$.

Chapter written by Massimo VANZI, Giovanna MURA, Laurent BÉCHOU, Giulia MARCELLO and Valerio Sanna VALLE.

The key difference in this protocol is the strong recommendation to exchange the roles of laser current and voltage, that is, to measure $I(V_{ext})$ and $P_{OUT}(V_{ext})$.

– The reason for using voltage, instead of current, as the driving quantity, is that a linear current ramp, for a case as in Figure 5.1, would reasonably apply current steps not smaller than 0.1 mA. This skips the whole micro- and nanoampere ranges, which carry important information about the current confinement mechanisms and the leakage phenomena.

– The voltage steps in the linear voltage ramp, again referring to cases similar to Figure 5.1, should be in the order of 5–10 mV.

– It should be evident that, no matter the current or voltage ramp as the driving quantity, the same *LIV* plot is obtained.

5.2. Voltage driving: the *logLIV* plot

The advantage of the exchange of roles between current and voltage can be appreciated in the alternative *LIV* plot that we recommend. It draws $I(V_{ext})$, $P_{OUT}(V_{ext})$ jointly in logarithmic scale (Figure 5.2), indicated as *logLIV*.

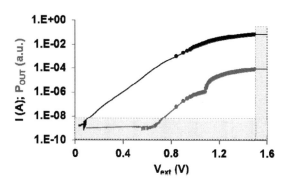

Figure 5.2. LogLIV – *The logarithmic version of the* LIV *curve. For a color version of this figure, see www.iste.co.uk/vanzi/reliability.zip*

5.3. Removing bad data: current compliance and ambient photocurrent

In Figure 5.2, we point out several features that frequently occur in real measurements:

– the choice to refer both current I and optical power P_{OUT} to the same axis will become clear: it is indeed possible to measure a conversion factor that transforms P_{OUT} into the radiative current I_{ph};

– the dots represent the measurement points that one would get on the same device if, instead of applying a linear voltage ramp (thin solid lines), as proposed here, a rough current ramp with a step of 1 mA is used, as usual;

– a vertical gray box indicates measurements that reached the compliance preset for the current meter. They are easily recognizable by directly inspecting the final raw data, and must be removed from the dataset;

– a horizontal box is indicated that shows a very small constant optical power. It simply means that the measurements have been performed when some ambient light was present. Data in this range must be removed as well, including from the current I, because the laser diode itself, and not only the power meter, is likely to have been perturbed by the ambient light.

We can now plot the clean version of the *logLIV* curves (Figure 5.3)

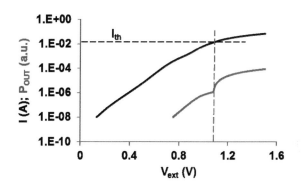

Figure 5.3. LogLIV *curves after removing the perturbed data. For a color version of this figure, see www.iste.co.uk/vanzi/reliability.zip*

5.4. Calculating internal threshold voltage V_{th} and series resistance R_S: the *logLIV* curves with respect to the internal voltage V

If we calculate the dV_{ext}/dI curve and plot it together with I and P_{OUT}, we get Figure 5.4.

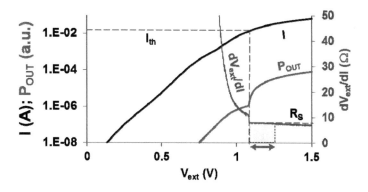

Figure 5.4. *The dV$_{ext}$/dI plotted together with I and P$_{OUT}$. For a color version of this figure, see www.iste.co.uk/vanzi/reliability.zip*

In Figure 5.4, the transition at the threshold is dramatically evident (vertical dashed line), and it is justified by equation [4.58] and Figure 4.18, in Chapter 4. The reason for repeating this image here is because of the double-arrowed voltage range under the highlighted box. This is the range where the experimental curve is close to the constant value indicated as R_S, the series resistance (which has a first estimate). In this range, we can suppose that the experimental pairs $(V_{ext,i}, I_i)$ are linked by the ideal equation:

$$V_{ext,i} = V_{th} + R_S I_i \qquad\qquad [5.1]$$

It is then possible to calculate the best values for the constants V_{th} and R_S beyond the fair accuracy of their graphical determination. For the set of data in the previous images, we get

$$V_{th} = 0.979\,\text{V}$$
$$R_S = 7.70\,\Omega$$

It is now also possible to replot the *logLIV* curves as a function of the internal voltage $V = V_{ext} - R_S I$.

The experimental curves recover the ideal behavior, with an evident clamp of the internal voltage at $V = V_{th}$. It should be noted that calculations are only introduced the value of R_S, and that the threshold voltage comes out graphically, in perfect ιgreement with its calculated value. The highest values of both curves, when plotted versus the internal voltage V, shifts to the left. This has been explained in Chapter 4 n section 4.9. Non-idealities and Figure 4.21, explaining the widening of the εmission area that reaches the voltage clamp after the main threshold transition.

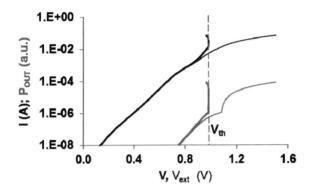

Figure 5.5. *The* logLIV *curves plotted with respect to the external voltage* V_{th} *(thin lines) and the internal voltage V (bold lines). For a color version of this figure, see www.iste.co.uk/vanzi/reliability.zip*

The immediate subthreshold region, when referred to V, behaves as an ideal diode, displaying a straight line as its logarithmic $I(V)$ characteristics. This will be nvestigated in more detail after the upscaling described in the next section.

5.5. Canonical logLIV: upscaling P_{OUT} to I_{ph}

5.5.1. Calculating I_{th} and reconstructing P_{TOT} and I_{ph} from P_{OUT}: quantum efficiency η_q

In the same above-threshold range that led to the internal voltage representation of the *logLIV* curves, another linear relationship links the laser current I and the measured optical output power P_{OUT}

$$I_i = I_{th} + KP_{OUT,i} \tag{5.2}$$

Equation [5.2] explains that the measured optical power P_{OUT} is always proportional to the total optical power P_{TOT} (equation [4.53], in Chapter 4), which, in turn, is proportional to the photonic current I_{ph}, which is the fraction of the total current I that is completely transformed into light (equation [4.42], in Chapter 4). Finally, the current I_{ph}, expressed as the difference between the total current I and the non-radiative current I_{nr} (equation [4.48], in Chapter 4), is identical to the difference $I - I_{th}$ in the above-threshold range, where I_{nr} is blocked at the value I_{th}.

In summary, if we find the value of the constant K in equation [5.2], we get

$$I_{ph} = KP_{OUT}$$
[5.3]

This allows us to evaluate the total power P_{TOT} by recalling equation [4.53], in Chapter 4

$$P_{TOT} = \frac{hv_{peak}}{q} I_{ph} = \frac{hv_{peak}}{q} KP_{OUT}$$
[5.4]

It is not trivial to realize that the multiplying factor hv_{peak}/q in equation [5.4] is the peak photon energy expressed in eV which, for semiconductor laser diodes whose peak emission is close to the bandgap amplitude, is a number close to unity (0.952 eV for the case in our example).

This means that measuring I_{ph} is also an appreciable numerical estimate of P_{OUT}.

Going back to equation [5.2], the dataset of our examples, calculated on the same range of the previous case, led to

$$I_{th} = 12.8 mA$$
$$K = 658 A/W$$

It is therefore possible to draw what we could define as the *canonical logLIV* (Figure 5.6), where the current I_{ph} is plotted, instead of the output power.

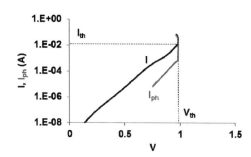

Figure 5.6. *The canonical logLIV, showing the total current I and the radiative current I_{ph} For a color version of this figure, see www.iste.co.uk/vanzi/reliability.zip*

5.6. Subthreshold Shockley parameters for I_{ph}: saturation current I_{ph0}, ideality factor n and quantum efficiency η_q

Figure 5.6 is worth several considerations:

– First of all, I_{ph} displays a sharp transition at the threshold, different from the smoother behavior of its theoretical formula. Moreover, the numerical fit of the subthreshold branch gives

$$I_{ph0} = 1.2 \times 10^{-12} A$$
$$n = 1.85$$

A simple way to properly fit the subthreshold branch of I_{ph} is to consider its Shockley-like characteristics, as given by equation [4.45], in Chapter 4, that is, conveniently written in the logarithmic form:

$$\ln I_{ph} = \ln I_{ph0} + \frac{qV}{nkT}, \qquad [5.5]$$

where the possibility for a non-unitary ideality factor n has been considered. This is a linear expression where all $\ln I_{ph}$ and all V values are known, which allows for calculating the best values for the unknown constants $\ln I_{ph0}$ and q/nkT, from which I_{ph0} and n are obtained.

Both the sharp transition and the non-unitary ideality factor strongly address the discussion of the non-idealities in Chapter 4.

– The radiative component I_{ph} is now plotted at the correct proportion with respect to the total current I. Referring to the current values measured close to V_{th} in the subthreshold range, we find, for our case, $I_{ph}/I = 0.065$. This is a *measurement of the quantum efficiency* η_q, as defined in equation [4.49], in Chapter 4.

– In the whole subthreshold range $V < V_{th}$, it is much smaller than I, making the attempt to plot the difference of $I - I_{ph}$ useless, which should not graphically differ from the plot of I.

– If we divide I_{ph} by the quantum efficiency η_q (in our case, this means to multiply it by about 15), we realize that the total current I, approaching the threshold, behaves as I_{ph}. This means that there is a range (the mA range up to I_{th} in our case, and, in general, in all cases where the threshold current is in the order of several mA) in which all currents are merely proportional to each other (highlighted region in Figure 5.7).

– The region highlighted in Figure 5.7 identifies that current and voltage ranges at which the lateral current confinement made the currents in the active region dominant.

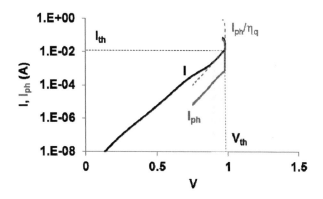

Figure 5.7. *The* canonical logLIV *and the upscaled I_{ph} (dashed line).*
For a color version of this figure, see www.iste.co.uk/vanzi/reliability.zip

5.7. Lateral current I_W: current confinement

It is also possible to identify the contribution of the lateral current, as highlighted in Figure 5.8. For completeness, and for comparison with the other data, the best fit was obtained by setting the saturation current I_{WS}, the ideality factor n and the total lateral resistance R_W, in the model of section 4.14, appendix E of Chapter 4, at

$$I_{WS} = 7 \times 10^{-10} \, A$$
$$n = 2.14$$
$$R_W = 155\Omega$$

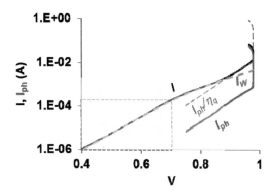

Figure 5.8. *Identification of the lateral current I_W (dashed thick line, mostly overlapped to I). For a color version of this figure, see www.iste.co.uk/vanzi/reliability.zip*

For I_W, one can apply the same procedure employed for the numerical fitting of the subthreshold branch of I_{ph}. Referring to the region highlighted in Figure 5.8, the Shockley-like behavior strictly belongs to the smaller boxed region, where the logarithmic curve is a straight line. Here, it is possible to calculate I_{WS} and n. The sole side resistance R_W then remains unknown in equation [4.98], in Chapter 4, and it can be calculated by extending the fit to the bending branch of the highlighted region. No relevant resistive shunting paths (current I_{sh} in equation [4.59] and Figure 4.20, in Chapter 4) have been detected.

5.8. Transparency voltage V_{tr} for peak emission and zero-loss threshold current I_{th0}

5.8.1. The loss-absorption ratio α_T / g_m

The knowledge of the peak emission energy (0.952 eV, as reported before) and the measurement of the threshold voltage V_{th} = 0.979 eV allows us to calculate the ratio α_T / g_m between total losses and absorption coefficient, by means of equation [4.31], in Chapter 4.

$$V_{th} = h\nu_{peak} + 2kT \ln \left(\frac{1 + \dfrac{\alpha_T}{g_m}}{1 - \dfrac{\alpha_T}{g_m}} \right)$$ [5.6]

This results in $\alpha_T / g_m = 0.25$.

The value $V_{tr} = h\nu_{peak} / q = 0.952$ V indicates the internal voltage at which the active region is transparent for the peak emission. The corresponding current is the threshold current at zero losses I_{th0}, as defined in equation [4.50], in Chapter 4. This voltage belongs to the area highlighted in Figure 5.7, where the sole currents of the active area are relevant. This means that one can add V_{tr} and I_{th0} as a coupled pair in the measured $I(V)$ characteristics (Figure 5.9).

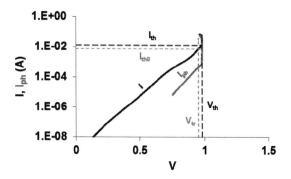

Figure 5.9. *The transparency condition displayed within the* canonical logLIV. *For a color version of this figure, see www.iste.co.uk/vanzi/reliability.zip*

5.8.2. *Application to diagnostics*

The proposed general protocol is a suitable tool suitable for *characterizing an* unknown device, and it is also a necessary step for starting *diagnostics*.

We can never stress the importance of starting any failure analysis with th characterization of a reference device enough.

Characterization means the complete reconstruction of the technology and th performances of a working device. The batch production of solid-state electron and electro-optical devices and their qualification protocols usually make each device ii a qualified lot identical to each other. This means that even a field return may hav a reference to be compared with.

Comparison, in turn, is the standard way for detecting changes. Mainly fo initial stages of degradation, a detected modification of a measurable quantity ma efficiently address the mandatory physical inspection.

Having the characterization of the initial (or reference) state available diagnostics may proceed at a faster pace: it is not necessary to repeat the whol characterization procedure for the degraded state, and only changes will be wort specific attention.

In order to show the protocol working on a real case of progressive degradatior we will focus on an experiment, detailed in Vanzi *et al.* (2013), that consisted of first, the creation, and then the enlargement, of a trench parallel to the active regio of a distributed feedback (DFB) InP-based laser diode tuned to 1310 nm (Figur 5.10). The aim was to modify the internal optical losses without changing the mirro properties or the performances of the active region.

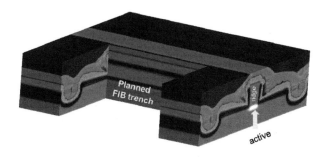

Figure 5.10. *The planned experiment of opening a trench at side of the active region (Vanzi et al. 2013). For a color version of this figure, see www.iste.co.uk/vanzi/reliability.zip*

The experiment in itself is not so representative of real degradation occurring during the operational lifetime of such devices. However, it has the advantage of knowing in advance what physically occurred, which should limit the possibility of imaginative interpretation, which is often misleading.

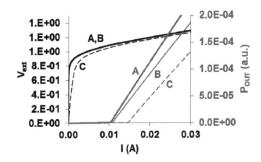

Figure 5.11. *The transparency condition displayed within the* canonical logLIV. *For a color version of this figure, see www.iste.co.uk/vanzi/reliability.zip*

Figure 5.11 reports the standard *LIV* of the initial state (A), the effect of a first trench drilled into the specimen (B) and then a relevant extension of that same trench (C).

We should note the displayed range for the laser current. It extends from 0 to no more than three times the initial I_{th}. It has been a recommended practical choice which can be appreciated, as shown in Figure 5.12. On the one hand, the smaller current range shown in Figure 5.11 allows a slight increase in I_{th} between the initial state A and state B, which is undetectable, as shown in Figure 5.12.

In Figure 5.12, one could conclude that the threshold current did not change which opens the door to a totally wrong interpretation (the only way to change the slope of P_{OUT} keeping I_{th} fixed is to change the optical coupling between the light emitter and its detector, which is a packaging problem, instead of what we know to be a physical alteration of the chip itself).

On the other hand, the linearity displayed by P_{OUT} at the beginning of the above-threshold range ends at higher current. This is a normal effect due to high injection levels (for which the model in Chapter 4 is no more accurate), which, in the author's experience, have no specific relevance in diagnostics, meaning that no known degradation mechanism is able to modify the sole nonlinear part of the power-current branch of the *LIV* curves without any evidence in the near-threshold domain.

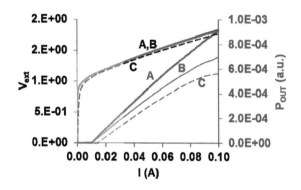

Figure 5.12. *The same LIV plot at an extended current range. The highlighted region corresponds to the range displayed in Figure 5.11. For a color version of this figure, see www.iste.co.uk/vanzi/reliability.zip*

As shown in Figure 5.11, we can apply the previous methods and, playing with the linear part of the above-threshold ranges, calculate the relevant constants.

	A	B	C
V_{th} (V)	0.9952	0.9968	0.9794
I_{th} (A)	0.0101	0.0105	0.0144
R_S (W)	9.9	9.8	9.5
K	87	103	117
η_q	0.012	0.010	0.009

Table 5.1. *Relevant constants for the initial (A) and degraded (B, C) states*

The *logLIV* plot points out other interesting details in the initial form, referring to the external voltage V_{th} (Figure 5.13) or to the internal voltage V (Figure 5.14). The *canonical logLIV* plot is shown in Figure 5.15.

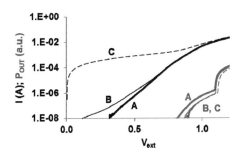

Figure 5.13. *The logLIV plot of the three states A, B and C referred to the external voltage V_{th}. For a color version of this figure, see www.iste.co.uk/vanzi/reliability.zip*

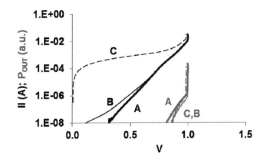

Figure 5.14. *The logLIV plot of the three states A, B and C referred to the internal voltage V. For a color version of this figure, see www.iste.co.uk/vanzi/reliability.zip*

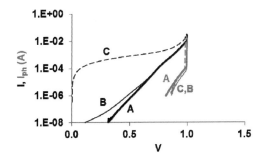

Figure 5.15. *Canonical logLIV with reconstruction of the radiative current I_{ph}. For a color version of this figure, see www.iste.co.uk/vanzi/reliability.zip*

The interpretation is particularly interesting, because it starts with a straight se of conclusions, but ends with an example of a logical trap, whose role here is to warn against the attempt to keep any modeling valid, in any case.

Let us start with the comparison of the initial state A and the first induced degradation state B.

All elements indicate the following points:

– Internal losses α_T slightly increased:

 - *this justifies the shift of the threshold current and the decrease in the quantum efficiency because of the decrease in I_{ph}.*

– No defects have been introduced affecting the active region:

 - *the total current I near the threshold did not increase.*

– A lateral conduction has been introduced:

 - *this results from the extra current at low voltage that appears in the B state and that should be compared with the shunting current I_{sh} as plotted in Figure 4.20 in Chapter 4.*

All these elements perfectly agree with the physical action of drilling a trench close to, but not inside, the active region. The electrical properties of the latter are no affected, but light escape is enhanced, thereby modifying the total losses. The wall of the trench introduce a leakage path that, for state B, does not affect the threshold.

It is interesting now to look at state C, which is a simple extension of case B Here, the threshold voltage decreases, which indicates a *reduction* of the losses, and there is also a decrease in the slope of the *LIV* curve, which is contrary. The serie resistance also decreased.

The reason is simple: the ideal electrical model does not perform. The shunting path has become so leaky as to drive a current comparable or larger than the threshold current I_{th}, as shown by the highly increased total current I in all the previous plots. The key point is that now there is a current path that is no longer parallel to the sole diode elements of the laser, (jointly indicated by block L in Figure 5.16) but is now parallel to the branch $R_S + L$.

A detailed analysis of this situation is reported in Mura *et al.* (2014).

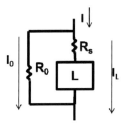

Figure 5.16. *Equivalent circuit for case C*

5.9. Graphical interpretation of changes in DC characteristics

In this section, we discuss the total equivalent circuit of a laser diode, as introduced in Chapter 4, and shown here in Figure 5.17.

We should keep in mind that:

– resistors R_S and R_{sh} behave as ideal resistors;

– the current I_{nr} is a standard Shockley current (equation [4.47], in Chapter 4);

– the current I_{ph} is the radiative current, well described by two functions (non-deality C: non-unitary ideality factor, in Chapter 4) and each represented by the ideal equation [4.44], in Chapter 4, with a different ideality factor n and the saturation current of the component with $n = 1$ much smaller than the component with $n > 1$;

– the lateral current I_W flows across a transmission line of resistors and diodes (as described and mathematically solved in section 4.14, appendix E of Chapter 4) whose relevant parameters are the total lateral resistance R_W and the total saturation current of the lateral diodes. For this reason, the schematics indicate the parallel as a simple resistor–diode pair, but the transfer characteristics are those given in section 4.14, appendix E of Chapter 4.

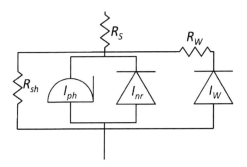

Figure 5.17. *Equivalent circuit of a laser diode*

It is now possible to draw (Table 5.2) an atlas of the predicted effects of changes in the various elements that describe the circuit.

It has been designed by comparing the *LIV*, *logLIV* and *canonical logLIV* plots for each case. The intensity of light emission has been referred to the externally measured optical power, with an arbitrary coupling constant, for the *LIV* plot. For both the *logLIV* and the *canonical logLIV*, it was preferable to plot the optical emission in terms of the radiative current I_{ph}, whose values are absolute. It follows that the second and the third plot in each line only differ for their abscissa: the external voltage V_{ext} for *logLIV*, and the internal voltage V for the *canonical logLIV*.

Parameter	Description	Starting value	Modified values
R_S	Series resistance	3 Ω	5 Ω, 10 Ω, 15 Ω, 20 Ω, 25 Ω
R_W	Total lateral resistance	100 kΩ	10 kΩ, 1 kΩ, 100 Ω, 1 Ω, 0.1 Ω
R_{sh}	Shunt resistance	100 GΩ	10 GΩ, 1 GΩ, 100 MΩ, 10 MΩ, 1 MΩ

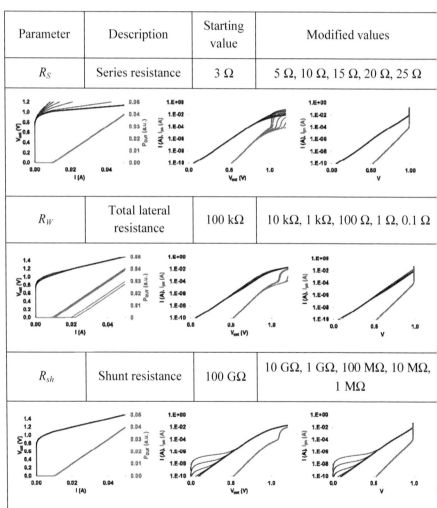

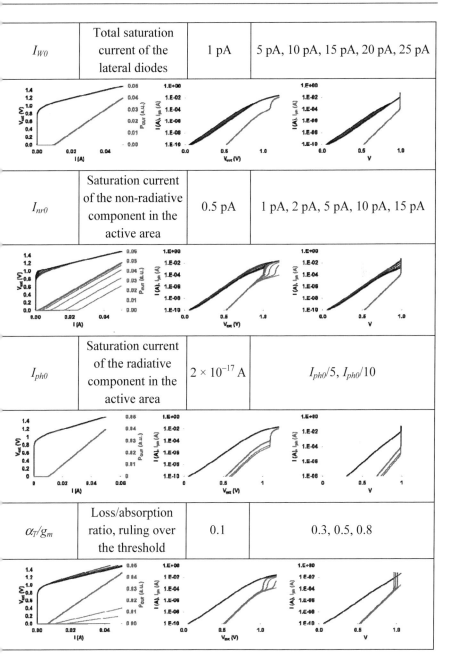

I_{W0}	Total saturation current of the lateral diodes	1 pA	5 pA, 10 pA, 15 pA, 20 pA, 25 pA
I_{nr0}	Saturation current of the non-radiative component in the active area	0.5 pA	1 pA, 2 pA, 5 pA, 10 pA, 15 pA
I_{ph0}	Saturation current of the radiative component in the active area	2×10^{-17} A	$I_{ph0}/5$, $I_{ph0}/10$
α_T/g_m	Loss/absorption ratio, ruling over the threshold	0.1	0.3, 0.5, 0.8

Table 5.2. *Atlas of LIV curves, varying a single parameter. For a color version of this table, see www.iste.co.uk/vanzi/reliability.zip*

There is not necessarily a link between the changes in Table 5.2 and physical degradation mechanisms. It was decided *not* to include variations of the ideality factors of the various diodes because of physical reasons: no mechanism is known that is able to impact that specific parameter. The changes in lateral resistance R_W could also have been excluded for the same reason. They were included for the completeness of the current-related changes.

Real degradation mechanisms rarely affect a single parameter. However, it is interesting to observe as the sole standard *LIV* plots are not able to distinguish between changes in different parameters, such as R_{sh}, I_{w0} and I_{ph0}, which are related to leakage paths, lateral current confinement and quantum efficiency, respectively. On the other hand, the triple plot *LIV*, *logLIV* and *canonical logLIV* is perfectly able to distinguish each case from the others.

It is impossible here to consider all possible degradation mechanisms (see Chapter 1) and their impact on possibly more than one parameter, which would require us to draw multiple combinations of the previous triple plots.

Here, we consider a single example, representative of the most dramatic and frequent case in laser diode degradation: the appearance of defects inside the active region. This is the case, not only of the catastrophical optical damage (COD), but also of the dark line defects in GaAs devices, and, in general, of all those cases where new defects are created inside the optically active region without connecting the p and sides of the diodes. The effect is twofold: on the one hand, the saturation current of the involved diodes increases, due to the average reduction of carrier lifetime, which corresponds to a higher recombination rate that is mostly non radiative. On the other hand, defects themselves become killers for photons, and then increase the internal optical losses.

Let us calculate the curves for a case in which (for a COD) the saturation current of the non-radiative component of the active region I_{nr0} and the total loss coefficient a_T increases, whereas the saturation current of the radiative component I_{ph} decreases.

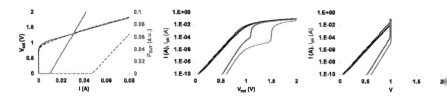

Figure 5.18. *Triple LIV plots for the case of defects grown inside the active region. For a color version of this figure, see www.iste.co.uk/vanzi/reliability.zip*

In Figure 5.18, the lines in the standard *LIV* plot have been kept thin, with the modified state indicated by dashed lines, in order to appreciate the most puzzling features in this case: the crossing of the IV characteristics, which is the combined effect of the indicated variations, cannot be explained differently.

5.10. Gain measurements

5.10.1. *Non-resonating optical cavities*

Protocols for measuring laser diodes would not be complete without mentioning spectral measurements, and, in particular, optical gain and losses as spectral functions. The model developed and presented in Chapter 4 is intrinsically spectral. It proceeds to spectral integration because of its focus on currents and total optical power. In principle, it could provide protocols for measuring spectral gain g, absorption coefficient g_m and loss coefficient a_T. And indeed, it does this, provided wave effects, and, in particular, optical resonances in Fabry-Perot cavities, or frequency selection, as in single frequency devices, can be neglected.

This approach to gaining measurement is different from the most popular ones, and even ignores the historical evolution, which started in 1973, from the Hakki–Paoli method (HPm) (Hakki and Paoli 1973, 1975) for Fabry–Perot (FP) cavities and included important refinements such as those introduced by Cassidy (1984). Even the authors moved from variants to HPm (Vanzi et al. 2016, 2017, 2018) which aimed to extend its applicability to DFB devices and to remove the need for an independent measurement of optical losses in order to measure gain. It was only at the end of this evolution that the problem of gain in non-resonating cavities was considered (Vanzi et al. 2019). Here, we reversed the sequence – we first investigated the no-loss devices and then introduced the FP resonances and the (modified) HP method.

If we had acquired a spectrum in the subthreshold range (where all emission lines are still competing to achieve the threshold that will lead one single frequency to dominate), we could describe it in terms of total emitted spectral power $|F_0|^2$ such as the integrand that appears in equation [4.40], in Chapter 4 (the reason for introducing the expression $|F_0|^2$ instead of a more intuitive P_v will become clear in the following equation):

$$|F_0|^2 = Vol \cdot hv \cdot \frac{\phi_v - \phi_{0v}}{\tau_C} \tag{5.7}$$

This can be referred to as gain by means of equations [4.39] and [4.26], in Chapter 4

$$
\begin{aligned}
|F_0|^2 &= W \left(1 + \frac{g}{g_m} \right)^2 \frac{\alpha_T g_m}{\alpha_T - g} \\
W &= \left[Vol \cdot c \cdot h\nu \cdot \frac{A}{4B} \right]
\end{aligned}
\qquad [5.8]
$$

where the coefficient W, which includes the frequency ν, both explicitly and also implicitly inside the Einstein coefficients A and B, can be considered nearly constant within the linewidth of the laser spectrum, and can be calculated once the volume Vol is estimated. Gain g, absorption coefficient g_m and loss coefficient α_T all being functions of photon frequency, but only g changing with the injection level, it should be sufficient to acquire at least two spectra at different currents I for measuring, a each frequency, the ratios g/g_m and α_T/g_m.

Decoupling of the above ratios can be achieved by considering a second gain equation, equation [4.33], in Chapter 4, written as

$$
H = \frac{g}{g_m} = \frac{1 - e^{\frac{h\nu - qV}{2kT}}}{1 + e^{\frac{h\nu - qV}{2kT}}}
\qquad [5.9]
$$

The internal voltage V is that same quantity that has been introduced for the $logLIV$ abscissa, and therefore each spectrum has its own H spectral function.

Equations [5.8] and [5.9] can be combined to obtain

$$
\frac{1}{g_m} + \frac{1}{\alpha_T} H_i - W \frac{(1 + H_i)^2}{|F_{0,i}|^2}
\qquad [5.10]
$$

where the suffix i has been introduced to indicate which (spectral) quantities are measurable from each spectrum. It should be clear that, knowing W, a minimum of two spectra allows us to solve a linear system for both $1/g_m$ and $1/\alpha_T$. With more spectra, the same result may be achieved with even better accuracy, i.e. by the least square method in section 5.11, the appendix. The key point is the reliable evaluation of W.

The example shown in Vanzi *et al.* (2019) starts from several spectra obtained from a DFB edge emitter with one facet of the cavity coated with an anti-reflection (AR) coating.

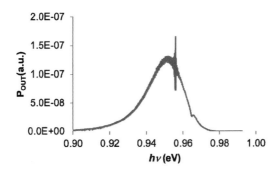

Figure 5.19. *Experimental subthreshold spectrum of a DFB edge emitter with AR facets*

The DFB peak has been numerically removed for the demonstration of the method, replacing it with the baseline interpolated from the surrounding spectrum. The result is plotted in Figure 5.20.

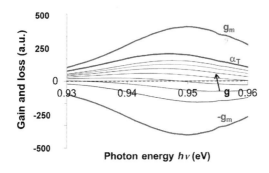

Figure 5.20. *Gain g, loss α_T and absorption coefficient g_m calculated from spectra, as in Figure 5.19. For a color version of this figure, see www.iste.co.uk/vanzi/reliability.zip*

The vertical scale has been left in arbitrary units, because of the problem of evaluating W. Here, the problem is that equations [5.7] and [5.8] refer to the *total spectral emission*, while the spectrometer has collected and converted only part of the emitted light: the term W includes an unknown factor which represents the collection and conversion efficiency of the equipment.

This has been solved in Vanzi *et al.* (2019) by applying the same upscaling that allows us to change the integral P_{OUT} into the total integral emission P_{TOT}, or to convert it to the radiative current I_{ph} (equation [5.4]).

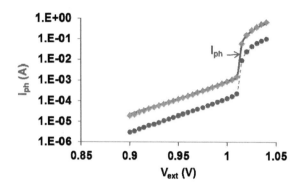

Figure 5.21. *Current I_{ph} (bold line) and the set of integrals of the various spectra (dots). Diamonds are the same spectra integrals after suitable upscaling to show their perfect alignment onto the I_{ph} plot. For a color version of this figure, see www. iste.co.uk/vanzi/reliability.zip*

Figure 5.21 demonstrates that it is possible to find a suitable factor for multiplying all spectra in order to make their integrals overlapped to I_{ph}. Therefore the simple transformation in equation [5.4] leads from I_{ph} to P_{TOT}.

In this way, the upscaled spectra are correctly represented by equations [5.7] and [5.8].

The result, for the given case, is that the vertical scale in Figure 5.20 coincides with the numerical gain, and loss values are expressed in cm^{-1}.

5.10.2. *Fabry–Perot and DFB cavities*

The problem of the absolute calibration of spectrum intensity disappears when we deal with resonating cavities. In this case, the HPm (Hakki and Paoli 1973, 1975) is a method based on ratios of intensities, and so eliminates any requirement for quantitative scale accuracy.

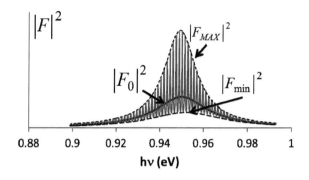

Figure 5.22. *A subthreshold spectrum as the modulation of the single-trip function $|F_0|^2$. The envelopes of maxima and minima are used in the Hakki–Paoli method for gain measurement*

The HP method starts calculating an expression for the subthreshold emitted spectral optical power $|F|^2$ (Figure 5.22) in the form of equation [5.11]

$$|F|^2 = \frac{|F_0|^2}{1+R^2 - 2R\cos\left(\dfrac{4\pi nL}{hc}hv\right)} \qquad [5.11]$$

where $|F_0|^2$ is the same as in the previous case and is indicated as the *single-trip* emitted spectral power, that is, the spectrum of the emission in case of no resonance null facet reflectivity). L is the length of the optical cavity, n is the refractive index, and the other basic quantities are as defined in Chapter 4. The term R collects gain and loss properties

$$R = \exp\left[-(\alpha_T - g)L\right] \qquad [5.12]$$

and simply vanishes in the case of no-reflection, bringing back $|F|^2$ to coincide with $F_0|^2$.

This is the ideal case of the infinite domain, and the real case of a device suitably coated on at least one of the boundaries of the optical cavity in order to have null reflectivity (which leads the mirror loss coefficient to α_m, and then the total loss to α_m infinite, see section 4.12, appendix C of Chapter 4, leading R to 0).

The method disregards the actual expression of $|F_0|^2$ and calculates the ratio r between the envelopes of maxima and minima:

$$|F|_{MAX}^2 = \frac{|F_0|^2}{\left[1-\exp\left(-\left(\alpha_T - g\right)L\right)\right]^2}$$

$$|F|_{\min}^2 = \frac{|F_0|^2}{\left[1+\exp\left(-\left(\alpha_T - g\right)L\right)\right]^2}$$

[5.13]

that is

$$r = \frac{|F_{MAX}|^2}{|F_{\min}|^2} = \frac{\left[1+\exp\left(-\left(\alpha_T - g\right)L\right)\right]^2}{\left[1-\exp\left(-\left(\alpha_T - g\right)L\right)\right]^2}$$

[5.14]

It is easy to insulate the argument of the exponentials, which we indicate as the spectral function S

$$S = \ln\left(\frac{\sqrt{r}+1}{\sqrt{r}-1}\right) = L\left(\alpha_T - g\right)$$

[5.15]

It is now possible to repeat the procedure (equation [5.10]), by combining the F function, equation [5.9], with the S function, equation [5.15]

$$S_i = L\alpha_T - Lg_m H_i$$

[5.16]

Here, the advantage is that one only needs to measure the length of the optical cavity L, without any further calibration.

One interesting point is that the method has been demonstrated (Vanzi et al. 2016, 2017) to also be applicable to DFB peaks, provided their result aligned with one of the FP maxima.

Limited to pure FP cavities, the problem of finding the envelopes of maxima and minima requested for calculating the ratios r, equation [5.14], has been simplified by substituting maxima and minima with a suitable integration over a spectrum period of the spectrum function and of its reciprocal (Vanzi et al. 2019).

Several studies (Vanzi *et al.* 2016, 2017, 2018, 2019) give details of the method for both the resonating and the non-resonating cavities.

Here, we comment on the underlying theory of this chapter. It seems, at first glance, to be based on the model in Chapter 4, which, in turn, looks quite different from the usual treatment of laser diode operation principles, and that still needs to be validated. This is just an apparent issue: gain measurement protocols are based on only three equations: the H function, equation [5.9], the $\left|F_0\right|^2$ spectral function, equation [5.8], and the Hakki–Paoli formula, equation [5.11]. The latter was formulated in 1973 (Hakki and Paoli 1973), but the first two are demonstrated, in Vanzi *et al.* (2019), to be derivable by Lasher and Stern's study in 1964 (Lasher and Stern 1964), as well as the concept of internal voltage, related to the separation of the quasi-Fermi levels, which is crucial for theory and practice, and was first proposed by Barnes and Paoli in 1976 (Barnes and Paoli 1976). Following this identification of the internal voltage, the H function demonstrated in Vanzi *et al.* (2016, 2017) is the same as that reported for gain theory, for instance, by Verdeyen (1995) in his classic book.

5.11. Appendix: a quick recall of the least squares method for simple cases

When experimental data pairs x_i, y_i are expected to be related by a linear relationship

$$y_i = a + bx_i \qquad [5.17]$$

one can estimate the values for the two constants a and b that lead to the best fit of the available experimental data to the given linear formula.

It is indeed obvious that if the data in equation [5.17] is not affected by measurement errors, we should have

$$a + bx_i - y_i = 0 \qquad [5.18]$$

In the presence of errors

$$a + bx_i - y_i = \varepsilon_i \qquad [5.19]$$

and ε_i is a random deviation from the null value and can be positive or negative.

This means that if we add up the squared deviations, they give a positive sum that should be as small as possible, in order to approach ideality.

$$\sum_i \left(a + bx_i - y_i\right)^2 = \min \tag{5.20}$$

In order to achieve the minimum, we allow the two unknown constants to vary, and then we calculate

$$\begin{cases} \dfrac{\partial}{\partial a} \sum_i \left(a + bx_i - y_i\right)^2 = 0 \\[2mm] \dfrac{\partial}{\partial b} \sum_i \left(a + bx_i - y_i\right)^2 = 0 \end{cases} \tag{5.21}$$

This leads to

$$\begin{cases} a \displaystyle\sum_i 1 + b \sum_i x_i = \sum_i y_i \\[2mm] a \displaystyle\sum_i x_i + b \sum_i x_i^2 = \sum_i x_i y_i \end{cases} \tag{5.22}$$

and then

$$\begin{cases} a = \dfrac{\displaystyle\sum_i y_i \sum_i x_i^2 - \sum_i x_i \sum_i x_i y_i}{\displaystyle\sum_i 1 \sum_i x_i^2 - \sum_i x_i \sum_i x_i} \\[6mm] b = \dfrac{\displaystyle\sum_i 1 \sum_i x_i y_i - \sum_i x_i \sum_i y_i}{\displaystyle\sum_i 1 \sum_i x_i^2 - \sum_i x_i \sum_i x_i} \end{cases} \tag{5.23}$$

All sums on the right-hand side of both equations [5.23] are calculated, and then a and b are also calculated.

It should be noted that equations [5.1], [5.2] and [5.5] in this chapter can be all handled by means of this simple least square approach.

5.12. References

Cassidy, D.T. (1984). Technique for measurement of the gain spectra of semiconductor diode lasers. *Journal of Applied Physics*, 56, 3096–3099.

Hakki, B.W. and Paoli, T.L. (1973). cw degradation at 300°K of GaAs double-heterostructure junction lasers. II. Electronic gain. *Journal of Applied Physics*, 44, 4113.

Hakki, B.W. and Paoli, T.L. (1975). Gain spectra in GaAs double–heterostructure injection lasers. *Journal of Applied Physics*, 46(3), 1299–1306.

Lasher, G. and Stern, F. (1964). Spontaneous and stimulated recombination in semiconductors. *Physical Review*, 133(2), A553–A563.

Mura, G., Vanzi, M. and Marcello, G. (2014). FIB-induced electro-optical alterations in a DFB InP laser diode. *Microelectronics Reliability*, 54.

Paoli, T.L. and Barnes, P.A. (1976). Saturation of the junction voltage in stripe-geometry (AlGa)As double-heterostructure junction lasers. *Applied Physics Letters*, 28(12), 714–716.

Vanzi, M., Mura, G., Marongiu, M. and Tomasi, T. (2013). Optical losses in single-mode laser diodes. *Microelectronics Reliability*, 53(9–11), 1529–1533.

Vanzi, M., Mura, G., Marcello, G., Béchou, L., Yannick, D., Le Gales, G. and Joly, S. (2016). Extended modal gain measurement in DFB laser diodes. *IEEE Photonics Technology Letters*.

Vanzi, M., Marcello, G., Mura, G., Le Galès, G., Joly, S., Deshayes, Y. and Béchou, L. (2017). Practical optical gain by an extended Hakki-Paoli method. *Microelectronics Reliability*, 76–77, 579–583.

Vanzi, M., Mura, G. and Martines, G. (2018). Further improvements of an extended Hakki-Paoli method. *Microelectronics Reliability*, 88–90, 859–863.

Vanzi, M., Mura, G., Rampulla, A., Marchetti, R., Sanna Valle, V. and Ueno, Y. (2019). Optical gain in laser diodes with null reflectivity. *Microelectronics Reliability*, 100–101, 113455.

Verdeyen, J.T. (1995). *Laser Electronics*, 3rd edition. Prentice Hall, Upper Saddle River.

Introduction to Appendix

The following pages are extracted from

Massimo Vanzi, "The Rules of the Rue Morgue," *ISTFA™ 1995: Conference Proceedings from the International Symposium for Testing and Failure Analysis*, ASM International, November 5–10, 1995, Santa Clara, USA.

I wish to personally thank ASM International for the reprint permission.

The *Rules* are mostly a *divertissement* about the logics of failure analysis.

It was written a long time ago, in 1995, when failure analysis (FA) was a widespread activity, carried out, not only by manufacturers, but, to a large extent, also by customers, and puzzling cases were usually shared within the electron device community. On the basis of my personal daily experience, at that time in a private telecom company, I was annoyed with the many cases of unsatisfactory FA reports and papers, and decided to analyze that feeling. I realized that the lack of logics was the most disturbing and recurrent problem. It ranged from the original sin of confusing failure mechanisms and failure modes (causes and effects), to the omission of some steps in the demonstration of the solution, or from the substitution of those logical and technical steps with someone else's opinion, or assumption from the beginning of an explanation and the adjustment or removal of physical evidence to demonstrate the solution.

I wanted to find a way to speak of that dissatisfaction, looking for a form that could be appealing, convincing and, why not, elegant. Just by chance, at that time, I happened to re-read *The Murders in the Rue Morgue* by Edgar Allan Poe. Everything was there! A puzzling event, a series of bad investigations, the logical

Introduction to Appendix written by Massimo VANZI.

analysis of the errors and the logical sequence that leads to a surprising solution, and the final demonstration.

I asked my friend Massimo Borelli, a skillful electron engineer, passionate teacher and clever painter, to illustrate the summary of the original tale, and then I complemented the job, trying to translate Poe's logics into a set of rules suitable for the failure analyst, and into a set of logical Violations that aim to help with understanding why and where an unsatisfactory FA report fails.

Over the (many) years since then, evidence of two things has piled up: a) violations never disappeared, so FA reports often continue to sound bad; b) more and more frequently, customers have entrusted failure analysis to the same manufacturers who could later be a counterparty in a technological or even legal dispute. This has preserved the *Rules* from obsolescence.

The *Rules* have continued to be applied, and periodically have been referred to, for over a decade, supported by plenty of new cases of questionable FA reports.

– Mura, G. and Vanzi, M. (2007). Failure Analysis of Failure Analyses: The Rules of the Rue Morgue, Ten Years Later. *IEEE TDMR*, 7(3).

– Mura, G. and Vanzi, M. (2016). Logics of Failure Analysis: 20 years of Rules of the Rue Morgue. *IEEE 23rd International Symposium on the Physical and Failure Analysis of Integrated Circuits (IPFA)*.

In the intervening years between writing the Rules and this book, I have been contacted by someone who was interested in reading the paper, and requested it, having found it difficult to find. For this reason, the original *Rules* are reprinted here. This has been faithfully reproduced and any errors that may appear are carried over from the original printing.

Appendix

The Rules of the Rue Morgue

"What caused the shift of the optical thresold of a laser diode, or *what place
a polysilicon whisker choose when it led the Iddq of your gate array to
increase* are puzzling questions"

(Massimo Vanzi, ISTFA95)

but

"What song the Syrens sang, or *what name Achilles assumed when he
hid himself among women,* although puzzling questions, are
NOT beyond all conjecture."

(Sir Thomas Browne, Urn Burial)
quoted by E.A.Poe at the beginning of
The Murders in the Rue Morgue

ANY FAILURE ANALYST with some practical experience remembers those
analyses that challenged his own professional reputation. They are those cases where
the novelty of the device or the puzzle of the symptoms put the analysis out of any
known possibility of solution. They are what we call *advanced failure analyses.*

Some time ago, asked to give a talk about "Strategies for Advanced Failure
Analysis", the Author had to answer the very fundamental question: "Does any
strategy exist for advanced f.a.?". Each new case is a different game, and its
occurrence in one or another field is mostly casual. Apart from skill, patience, luck

---

Appendix written by Massimo VANZI.

and memory, it seems that the only old good common sense may drive our steps through unexplored seas. But, does good sense have rules?

It was completely casual, in that same period, to find in a store a copy of *The Murders in the Rue Morgue* [1] by Edgar Allan Poe. And it was also a sort of Revelation! Not only it is the founder of the modern detective story (and how many times we call our analyses as *investigations*!), but even its preface clearly states how that tale aim to deal with the analytical faculties of the human brain, that is the capability of solving the most intricate problems by the sole observation and reasoning. Monsieur Dupin, the incredible protagonist of the tale, summarizes in some few pages the deductive method that will then be extensively used by Sir Arthur Conan Doyle for creating his celebrated son, Sherlock Holmes.

The reduced size of the tale, its declared thesis, its amusing comments on the most common errors of the *parisien police,* render it an extraordinary mirror for any detective's face. Failure Analysts will readily find how to write down its paraphrase by properly substituting for characters, situation and techniques.

This paper cannot avoid a brief summary of the story, with the aid of few appropriate drawings. Then, it will attempt to point out the Rules of the Rue Morgue, and to discuss them. Completeness should finally request for verification in practical cases, which should be easy by simply reading the many reported cases of advanced f.a. Modesty and prudence, on the other hand, will move the Author to limit his comments within his own scientific production. May be not a case that over his near 80 papers, only two seemed (to him) to completely obey the Rules of the Rue Morgue.

The Murders in the Rue Morgue (a summary)

The crime. Paris, last century. After having introduced Monsieur C.Auguste Dupin and its amazing deductive capabilities, the tale moves from three subsequent editions of the *Gazette des Tribunaux.*

The evening edition announces: EXTRORDINARY MURDERS. At three o' clock in the morning the inhabitants of the Rue Morgue "were aroused by a succession of terrific shrieks, issuing, apparently, from the fourth story of a house [...]known to be in the sole occupancy of one Madame L'Esplanaye, and hi daughter[...]"

"After some delay [...]the gateway was broken in with a crowbar, and eight or ten of the neighbours entered, accompained by two *gendarmes*"

"By this time the cries has ceased; but, as the party rushed up the first flight of stairs, two or more rough voices, in angry contention, were distinguished, and seemed to proceed from the upper part of the house. As the second land was reached, these sounds, also, had ceased, and everything remained perfectly quiet"

The door is forced open, and the apartment appears "in the wildest disorder - th furniture broken and thrown about in all directions[...]. On a chair lay a razo besmeared with blood[...], three long and thick tresses of grey human hair, als dabbled in blood[...]". On the floor, jewels, two bags with nearly four thousan francs in gold. " Of Madame L'Esplanaye no traces were here seen".

But, after a search was made in the chimney, "the corpse of the daughter, head downwards, was dragged therefrom; it having been thus forced up the narrow aperture for a considerable distance. The body was quite warm. [...] many excoriation were perceived[...]. Upon the face were many severe scratches, and, upon the throath, dark bruises, and deep indentation of fingernails, as if the deceased has throttled to death".

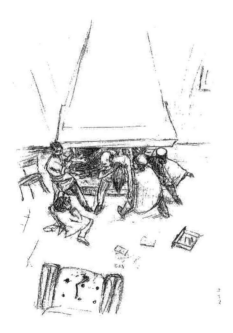

"After a thorough investigation of every portion of the house, without farther discovery, the party made its way into a small paved yard in the rear of the building, where lay the corpse of the old lady, with her throath so entirely cut that, upon an attempt to raise her, the head fell off. The body, as well as the head, was fearfully mutiled - the former so much so as scarcely to retain any semblance of humanity."

"The next day paper had [...] additional particulars". Two neighbours, *Pauline Dubourg,* laundress and *Pierre Moreau,* tobacconist, deposed that the old lady and the daughter seemed very affectionate towards each other. Lived a very retired life. They were in excellent pay. Madame L. was told to tell fortunes.

Other neighbours confirm that no one was spoken of as frequenting the house. The shutters of the front windows were seldom opened. Those in the rear were always closed, with the exception of a large back room.

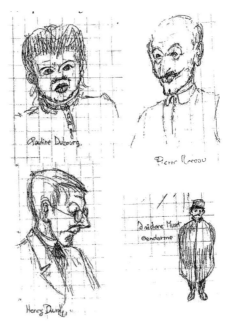

Isidore Muset, gendarme, who opened the door, confirms the chronicle of the events, and specifies that he "heard two voices in loud and angry contention - the one a gruff voice, the other much shriller - a very strange voice." The former was surely of a Frenchman (could distinguish the words *sacré* and *diable*). The other was the voice of a foreigner, maybe a Spanish.

Henri Duval, a silversmith, "corroborates the testimony of Musèt in general. [...] The shrill voice [...] was that of an Italian." He is not sure that it was a man's voice. "Was not acquainted with the Italian language."

_____ *Odenheimer, restaurateur*, not speaking French, is native of Amsterdam. The voices were both of Frenchmen.

Jules Mignaud, banker, says that Madame L., three days before her death, took out in person the sum of 4000 francs. A clerk was sent home with the money.

Adolphe Le Bon, the clerk, deposes that he accompanied Madame L. to her residence with the money, and then departed. He did not see any person in the street at the time.

William Bird, an Englishman, heard distinctly *sacré* and *mon Dieu.* The other voice appeared to be that of a German. He does not understand German.

Alfonzo Garcio, a Spanish undertaker, judges that the second voice was that of an Englishman, but he does not understand English, while *Alberto Montani*, an Italian confectioner, thinks it was the voice of a Russian (but he never conversed with a native of Russia).

Paul Dumas, physician, and *Alexandre Etienne*, surgeon, describe the horrible condition of both bodies. Only one or more very powerful men could be responsible for such a butchery.

"Four of the above-named witnesses[...] deposed that the door of the chamber in which was found the body of Mademoiselle L. was locked on the inside[...]The windows, both of the back and front room, were down and firmly fastened from within". No other possible accesses were discovered. The voices ceased few minutes before forcing the door, but no person was seen within the rooms.

The evening edition of the *Gazette*, reports that Adolphe Le Bon has been arrested and imprisoned for the murders.

About the method. Here Dupin comes into play. He is not satisfied of the progress of this affair: "We must not judge of the means [...] by this shell of examination. The Parisian police, so much extolled for *acumen*, are cunning, but n more. There is no method in their proceedings, beyond the method of the moment. They make a vaste parade of measures; but, not unfrequently, these are [...]ill adapted to the object proposed. [...] The results attained by them are not unfrequently surprising, but, for the most part, are brought about by simple diligence and activity. When these qualities are unavailing, their schemes fail. Vidocq, for example, was a good guesser, and a persevering man. But, without educated thought, he erred continually by the very intensity of his investigation. He impaired his vision by holding the object too close. He might see, perhaps, one or two points with unusual clearness, but, in doing so he, necessarily, lost sight of the matter as a whole".

The survey. Dupin obtains the permission for a personal inquiry, that starts with a long and detailed examination of the whole neighbourhood and the front and the rear of the house. Going upstairs, he may verify the exactness of the informations given by the *Gazette*: the room and the bodies of the victims are exactly as described. Nevertheless, "Dupin scrutinezed everything" again.

The Analysis. The crime, the voices, the empty rooms locked from inside are the conflicting elements of the problem. For the sake of the method, Dupin lists the only three possible directions to follow:

Ghosts. This could explain everything, but "neither of us believe in praeternatural events"

Homicide/suicide. This agrees with the locked rooms, but not with the mechanics of the murders nor, at least, with the male french voice.

Homicide by third persons. The most convincing hypothesis, leaving the mistery of the empty locked rooms.

Discarding the first two "possibilities" because of evidence, the third is taken into account as the *starting* point for analysis.

Dupin points straight towards the most *peculiar* aspects of the events. The key sentence is NOT "what has occurred" but "what has occured that has never occured before", because the police "had fallen in the gross but common error of confounding the unusual with the abstruse. But it is by these deviations from the plane of the ordinary, that reason feels its way, *if at all*, in its search for the true."

Accordingly, the most *peculiar* aspects are given by

1) the extraordinary violence of the crime;

2) the absence of any evident motive;

3) that strange shrill voice whose nationality none of the many witnesses was able to recognize.

The most *unexplicable* feature, on the other hand, is given by the empty locked rooms. Here a chain starts of questions whose answer causes a new question, up to a surprising possibility of solution:

1) Why the rooms are locked and empty? Because the killers escaped.

2) From where did they escape? The only possibilities are the door and the windows. From the door it is impossible, because the party was ascending the stairs, and the door was locked from inside. So, the assassins fled through a window.

3) How could the killers shut the window from inside and to lock it?

The windows must have a spring for automatic shutting, that has released at the flight of the criminals.

4) But both windows have a nail blocking them from inside. How is it possible? One of the nails *must* be defective, and here Dupin reveals one of the findings of his careful examination: one of the two nails is so rusty to be broken in the middle possibly since a long time.

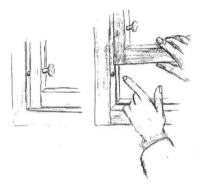

5) How could anybody leave from the window of the fourth storey of a building? "About five feet and a half from the casement in question there runs a lightning rod[...]" that could be reached from the window by jumping to and from the swinging shutter of the window.

6) Is it possible? Yes, provided the murderers have a near superhuman agility.

From the same way the criminals must have entered the apartment.

The peculiar aspects of the crime are updated by this new element: the culprit must diplay superior vigour and agility.

Here Dupin reveals his other two findings, obtained as a consequence of the above indications: a "little tuft from the rigidly clutched fingers of Madame L'Esplanaye" that is readily recognized as no *human* hair, and a drawing reproducing the impression of the fingers of the assassin on the throath of Mademoiselle L. It displays an hand much larger than a human hand!

The killer, Dupin reveals, is an Ourang-Outang of the East Indian Islands, that probably escaped from its owner.

The tale rapidly runs to its conclusion: a sailor is traced by means of a stratagem, and forced to confess. He got the Ourang-Outang during a voyage to the Indian Borneo, and decided to bring and sale it in Paris. It was kept locked into the sailor cabin on the ship, where it amused itself imitating any gesture of the sailor. The latter tried to stop the monkey when it happened to find his razor, and to mime a shave, but the beast, frightened by his reaction, succeeded in running away, followed by the sailor in despair. The run ended in the Rue Morgue, where the monkey, attracted by the light at the fourth storey of a buiding, raised the lightning-rod and entered the apartment of Madame L'Esplanaye.

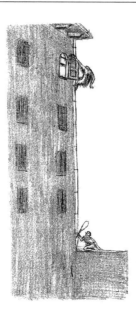

The sailor was able to follow the Ourang-Outang up to the summit of the rod, but not to reach the window: here he could do nothing else than cry when the monkey, frightened by the screamings of the two ladies, massacred them and then escaped. He also ran away, despaired for what had happened.

Of course, the original tale is much richer than this summary, and the style of that Author is uncomparably higher than that of the present one, who strongly recommends that version.

The Rules of the Rue Morgue

The logical rules underlying the Poe's tale are evident. Anyway, it could be *amusing* to attempt their translation into the world of Failure Analysis. This attempt only aims to be a proposal, open to any criticism, improvement or extension.

It is just for the sake of clarity that the definition of failure modes, failure mechanisms and failure analysis are here recalled:

– failure *modes* are the observable manifestations of a faulty state of a device;

– failure *mechanisms* are the physical causes of the observed failure modes.

– failure *analysis* is the process that leads from failures modes to failure mechanisms.

There is a common thought under any f.a: the failure mechanism may be discovered. There is also a widespread experience of reported f.a.'s that leave the reader somewhat unsatisfied. All these aspects are dealt with along the Poe's tale, so that it is possible to name the single items of the following list, according to the characters that are called into play for illustrating each of them.

The Dupin Postulate. Given a specific specimen, the global observation of the complete set of failure modes leads to the unique possible failure mechanism.

Corollary. When a subset of failure modes is sufficient to lead to a unique failure mechanism, any other failure mode must be explained by that mechanism.

The *Vidocq Violation.* Any F.A. that explains only a subset of the observed f. modes is defective or false.

The *Ghosts Violation.* Any F.A. that links f.modes to a non-mechanism is false.

The *Fool Lady Violation.* Any F.A. that individuates an impossible mechanism for the given specimen is false.

The *Le Bon Violation.* Any F.A. that relates f. modes and f. mechanism by means of non-deterministic links is defective or false.

Many violation cases could be cited.

In the Author's experience and opinion, terms as Burn-out, Catastrophic Damage, Dark Defect, etc. are all examples of Ghosts Violation: none of the indicated solutions is a failure mechanism, but, in the best case, just a tautology for one of the failure modes.

In the same way, EOS and ESD, that are perfectly legitimate failure mechanisms, are the most frequent findings under the Le Bon Violation, at least in the reports of some f.a. labs: a SEM photograph of a damaged (?) metallization, possibly not involved at all in the displayed electrical alteration, may be the only supplied "evidence". On the other hand, the large statistical incidence of those mechanisms (a sort of butler of any crime) may induce some relaxation in the investigative rigour.

The Fool Lady Violation takes place when the history and the nature of the specimen does not allow an otherwise legitimate mechanism to take place: overstress of unbiased specimens, corrosion in dry atmosphere (the Author has one of the oldest slides of his archive that shows a "clear" example of galvanic corrosion. Unfortunately, the specimen had been just overheathed, and the "corrosion" was the good old purple plague). On the other hand, because of the widespread unreliability of the specimen historical data supplied to the f.a. labs, a conclamated Fool Lady Violation may be a flag for investigating on the truth of the specimen anamnesis.

The Vidocq Violation is scarcely reported, but is probably quite common: any f.a. unable to give account of the whole set of observable failure modes may be made-up by just paying more attention to some informations than to others. The ultimate aspect, if unsatisfactory, is that of a Le Bon Violation, until the "complementary" findings come out.

To go beyond the phase of criticism, the Rules of the Rue Morgue for performing a f.a. may be summarized in few common sense sentences:

1) *The Survey Rule.* Collect in person as many informations and non-destructive measurements as possible.

The last is always a larger set than that leading to the first detection of the failure: nobody controls the i-v characteristics of an optoelectronic device during its operating life but, once the optical behavior fails, electrical informations should be collected.

2) *The Extraordinary Rule.* Identify anything peculiar within the set of available failure data and list, separately for each, its possible causes.

3) *The Inevitable Suspicion Rule.* By the above set of possible interpretations extract the most recurrent and mark, the contrary indications, if any.

4) *The Impossible Chain.* If any contrary indication exist, conflicting with the proposed interpretation, analyze the possible causes of that discrepancy *as a failure mode of the f.a.* by looping the Extraordinary and the Inevitable Suspicion Rules Add the partial interpretations, discovered along the chain, to the input data for the Survey Rule.

5) *The Orang-Outang Evidence Rule.* Once any main interpretation has been identified, search for other detectable manifestations that might be correlated with it Add to the input data for the Survey Rule.

6) *The Detective's Loop.* The above steps should be iterated until no new input data are obtained for the Survey Rule. At this point, three possibilities may occur:

a) A unique interpretation has been found, and it is a failure mechanism. This concludes the f.a.

b) A unique interpretation has been found, but it is not a failure mechanism. In this case an Impossible Chain should be attempted to exploit the relationship between this intermediate failure mode and a mechanism.

c) The loop was not able to find any unique interpretation. This should require the search for other data to be added to the first step, the Survey Rule.

Discussion

The Author is perfecly aware that nothing newer than usual procedures has been proposed. One could summarize the whole deal with the sentence: *you may or may not find an explaination for your problem.*

Anyway, it seems useful to recognize that f.a. is a difficult task and that, no matter how much you are urged to find an answer, each of the three above output of the Detective Loop may occur. In any case, it IS a result, that, for cases b and c, requires more objective data to be improved.

The same Poe's tale, that is a literary exercise, builds an ideal situation leading to a unique solution by means of circumstantial proofs that may not have significantly different solution. And, in any case, the Ourang-Outang Evidence is inserted from *outside* (it is not because of his analytical faculties that Dupin *finds* the red hairs of the beast).

When E.A.Poe attempted to face a real crime, with the tale *The Mistery of Marie Roget* [2], a declared sequel to *The Murders in the Rue Morgue,* he was able to guess the exact solution, but he did not complete the Detective's Loop, because of the lack of any Ourang-Outang Evidence. His solution was successively demonstrated to be correct and no violation could be detected of the Dupin Postulate, but, from the f.a. point of view, it is not complete.

On the much more limited experience of the present Author, two papers might be proposed to the reader's attention, both coming from ISTFA91: "Interpretation of Failure Analysis Results on ESD-Damaged InP Laser Diodes" [3]. and "Failure Analysis of CMOS devices with anomalous IDD currents"[4].

The first starts with sudden failures displaying typical long term degradation modes. Here an Impossible Chain starts, whose conclusion relates an high sensitivity to ESD to high power optical generation, local melting and recrystallization of the extremely narrow active layer, formation of extended defects with very high transversal confinement, alteration of the local carrier lifetime and reduction of optical gain, EBIC distortion, EL quenching and peculiar modification of the i-v characteristics, as observed at the first investigation stage.

The second is a sort of automation of the Inevitable Suspicion Rule: the Voltage Contrast images of high current absorbption logic states in a CMOS semicustom device are compared and the only recurrences are saved. It is a sort of verification of the Corollary to the Dupin Postulate: with the logical intersection of a limited number of independent informations, it is possible to point to the only gate, within hundreds of thousands, that is responsible for the detected failure mode. Here even the Ourang-Outang Evidence is supplied, by means of *a poteriori* liquid crystal analysis.

Conclusion

This paper, it is evident, is mostly a joke, based on the fascinating (but not original) consideration of any f.a. as a detective story. The Poe's tale is a perfect instrument (but surely not the only possible one) for playing this game.

Anyway, if any practical application of "the Rules of the Rue Morgue" may be expected, it is on the possibility of defining what leaves us unsatisfied when a f.a. result sounds out of tune. The reported Violations to the Dupin Postulate summarize the objections that the Author would like to repeat for his own analyses, and for those cases in which he is required to review the work of other Authors.

On the constructive side, the proposed Rules, it has been repeatedly said, are common sense indications, and are surely not exaustive, on a practical ground. Skill, patience, luck and memory are also required, but, unfortunately, not always and not together available.

It should be of the greatest aid for the Failure Analyst community, in any case that each public report could point out how it obeyed to a widely accepted set of f.a rules. Maybe - why not? - the Rules of the Rue Morgue.

As a last consideration, for concluding the joke, the Author invites his readers to open the original Poe's tale at the very beginning of the story, when Monsieur Dupin is introduced. Thinking at the Failure Analyst as a member of the excellent family of the Scientists, many of us will sigh and smile.

Acknowledgements

The Author is indebted with Karel Van Doorselaer for his encouragement in following the first idea of "The Rules of the Rue Morgue". He suggested a tale by Agatha Christie and proposed the exciting idea of the *lying devices,* that has not been exploited in this paper. This could be a stimulus for further amusing exercises.

The hidden second Author of this paper is Massimo Borelli, a clever teacher of Electronics, and a good old friend of the Author, that illustrated "The Murders in the

Rue Morgue" in such a fascinating way. Thanks to him, it seems that fantasy and semiconductors are not so far.

1] E.A.Poe: "The Murders in the Rue Morgue", Graham's Magazine, 1841.

2] E.A.Poe: "The Mistery of Marie Roget", Southern Library Messenger, 1842-3.

3] M.Vanzi, M.Giansante, L.Zazzetti, F.Magistrali, D.Sala, G.Salmini and F.Fantini: "Interpretation of Failure Analysis Results on ESD-Damaged InP Laser Diodes". Proc.17th International Symposium for Testing and Failure Analysis ISTFA/91, Los Angeles, November 1991, pp.305-314.

4] R.Bottini, D.Calvi, S.Gaviraghi, A.Haardt, D.Turrini and M.Vanzi:: "Failure Analysis of CMOS devices with anomalous IDD currents". Proc.17th International Symposium for Testing and Failure Analysis ISTFA/91, Los Angeles, November 1991, pp.381-388.

List of Authors

Laurent BÉCHOU
Laboratoire de l'Intégration du
Matériau au Système (IMS)
UMR-CNRS 5218
University of Bordeaux
France

Daniel T. CASSIDY
McMaster University
Ontario
Canada

Yves DANTO
Retired
Laboratoire de l'Intégration du
Matériau au Système (IMS)
UMR-CNRS 5218
University of Bordeaux
France

Yannick DESHAYES
Laboratoire de l'Intégration du
Matériau au Système (IMS)
UMR-CNRS 5218
University of Bordeaux
France

Mitsuo FUKUDA
Toyohashi University of Technology
Japan

Samuel K. K. LAM
IPG Photonics
Oxford
Massachusetts
USA

Giulia MARCELLO
STMicroelectronics
Agrate
Italy

Laurent MENDIZABAL
Département optique et photonique
(DOPT)
Service des Matériaux et
Technologies pour la Photonique
Laboratoire d'Électronique et de
Technologie de l'Information (LETI)
Commissariat à l'énergie atomique et
aux énergies alternatives (CEA)
Grenoble
France

Giovanna MURA
Department of Electric and
Electronic Engineering (DIEE)
University of Cagliari
Italy

Yves OUSTEN
Laboratoire de l'Intégration du
Matériau au Système (IMS)
UMR-CNRS 5218
University of Bordeaux
France

Valerio Sanna VALLE
Department of Electric and
Electronic Engineering (DIEE)
University of Cagliari
Italy

Massimo VANZI
Department of Electric and
Electronic Engineering (DIEE)
University of Cagliari
Italy

Frédéric VERDIER
Laboratoire de l'Intégration du
Matériau au Système (IMS)
UMR-CNRS 5218
University of Bordeaux
France

Index

A, B, C

absorption, 215, 223, 225–227
aging curve, 58, 59, 61, 63, 65, 66,
 68, 69, 71–74
 knee, 63, 72
annealing, 63–70, 72, 74
 optical defects, 64–67
 output efficiency, 64, 66, 67
black body, 142, 148, 149, 155
COD (catastrophic optical damage),
 26–31, 33–36, 48
 -induced higher order lateral
  mode, 33
 lateral mode induced, 35

D, E, F

defect (*see* growth rate)
 annihilation, 63, 64, 67, 68
 density, 55, 56, 62
 generation, 55, 64
 population, 52–56, 58, 59, 68, 74
degradation (*see* threshold)
 dynamics, 61
 fast, 61
 model, 51, 52
 saturable, 52–54, 58, 59, 61, 64, 72

slow, 61
 transient, 71
degradation mechanisms, 2–7, 12, 17,
 25, 42, 45, 46
ESD (electrostatic discharge), 26,
 36–43, 45, 46, 48
failure mechanisms, 82, 134

G, H, L

growth rate
 defect, 52, 54, 60, 62, 68, 74
 intrinsic, 55, 56
 population, 54, 56
Hakki–Paoli method, 225, 228, 229,
 231
laser diode, 81, 82, 84, 85, 92, 117,
 125, 127, 129, 130, 132–137
 degradation, 224
 diagnostics, 216, 218
 modeling, 139
least square, 226, 231, 232
lifetime, 79, 81–85, 94, 103, 126,
 132, 134–136
LIV plot, 207, 208, 218, 222, 224,
 225
 log, 208, 219, 222

logistic function, 56
loss/absorption, 215, 223, 225–227

M, O, P

maximum sustainable population, 56, 57, 68
Monte Carlo method, 79, 82–84, 86–89, 96, 101, 103, 112, 117, 118, 120–125, 133, 136, 137
multicomponent
  degradation curves, 61, 71
  model, 61
optical
  gain, 152, 158, 188
    measurement, 225, 229, 231, 233
  power, 139, 140, 151, 165, 167, 171
  systems, 1–3, 5, 11, 17, 43
optoelectronics, 79, 80, 84, 132, 135–137
  device, 79, 84, 132, 137
photon density, 143, 149–151, 153, 154, 156, 157, 160–162, 168, 183, 187, 190–193
physical insight, 56, 156

Q, R, S

quasi-Fermi levels, 139–141, 143, 146, 148, 150, 159, 168, 181, 182, 198
radiative current, 165, 170, 183
  non-, 58, 142, 143, 167–169, 176, 181, 199, 201

rate equation, 54, 56, 61, 64, 65, 68, 139–143, 150, 152, 159, 185
  for defect population, 54
reliability, 1–3, 15–18, 20–22, 26, 28, 29, 37, 41–45, 48, 79–81, 83, 94, 96, 101, 105–111, 117, 118, 121–126, 132, 134, 135–137
S-TEM, 25, 29, 31, 32, 34, 36–39, 42
semiconductor lasers, 12, 42, 43
statistics, 81, 83, 84, 86, 103, 117, 120, 126, 133
stimulated emission, 139, 149, 151, 152, 154, 159, 168, 188, 203

T, V, W

telecommunications, 79, 81, 82, 118, 132, 135–137
threshold
  current, 52–54, 58–60, 64, 66, 69, 213, 215, 218, 220
  degradation, 52, 53, 60, 62
  voltage, 209, 210, 215, 220
t^m time dependence, 58
Verhulst–Pearl equation, 51, 52, 56, 57, 74
wear-out, 60, 72–75